The Decline and Approaching Fall of the U.S.

When Social Security & Other Trust Funds Fail

R. Earl Hadady

First published by AuthorHouse 11/04/04

ISBN: 1-4184-6702-2 (e-book)
ISBN: 1-4184-4922-9 (Paperback)
ISBN: 1-4184-4921-0 (Dust Jacket)

Library of Congress Control Number: 2003096712

This book is printed on acid free paper.

Printed in the United States of America
Bloomington, IN

Other books by R. Earl Hadady

Contrary Opinion, How to Use It for Profit in Trading Commodity Futures

Opening Price Statistical Data on the Futures Markets

Historical Commodity Seasonal Charts

Historical Commodity Spread Charts

How Sick is Uncle Sam?

Contrary Opinion, Using Sentiment to Profit in the Futures Markets

Treason in High Places (a suspense novel)

Contrary Opinion, Using Sentiment to Profit in the Futures Markets (Chinese translation)

earl-hadady.com

for more information

Preface

My interest in the fiscal health of our government began a number of years ago. Like most Americans, Uncle Sam's finances were a mystery to me—the vastness and complexities seemed daunting—yet, I believed they could be put in understandable terms. My learning curve took longer than I'd like to admit and I took many wrong turns, but my findings were reducible to a few simple statements which you can easily verfy.

The *Monthly Treasury Statement of Receipt and Outlays* is almost a one-stop information center. Social Security and the Public Debt complete the sources for information—all data is available on the internet.

I was aware of the increasing public debt, but the President and Congress seemed to limit their concern to oration. While a few members of Congress and prominent men in and out of government voiced warnings, they were in a very small minority. There were even a few doomsayers among us who were saying our government was in serious fiscal trouble—but they provided no clear evidence and no timetable.

I narrowed down my understanding of Uncle Sam's fiscal condition to five keys.

The first key: Uncle Sam's doesn't adhere to the rules that he requires of private companies. The government throws all receipts it gets from taxes, etc. into a pot. All outlays are made from the same pot. This is called commingling of funds—it is illegal in private companies. Routinely when the government has insufficient funds in the pot to cover outlays, they borrow from you, me and others willing to load them money. Typically, foreign sources comprise almost 25% of the lenders—this could be a serious problem for the U.S. if these sources decide not to renew their loans.

The second key: The government's fiscal problems are centered in the trust funds—incidentally, they occupy only a half-page on page 30 of the 32 page *Monthly Treasury Statement of Receipt and Outlays.* Contrary to the small amount of space given to the trust funds, they constitute essentially 50% of all government receipts and outlays. By comparison, the Department of Defense gets only about 12% of the outlays. In short, the trust funds control the budget—the executive, legislative, and the judicial branches of government with their some 14 departments and 60 independent establishments and government corporations turn out to be incidental players.

The third key: The government budget surpluses and deficits are figured incorrectly and are very misleading. If you were told General Motors took money collected from their employees for retirement, health insurance, etc. and added it to car sales to determine their profit or loss for the year, you would most likely laugh—but that is exactly what the government does. Trust fund receipts are added to tax receipts etc. before outlays are deducted. The trust funds are almost entirely disability insurance, life and health premiums, retirement contributions, hospital insurance, old age and survivors insurance, supplementary medical insurance, unemployment insurance, etc. **The last surplus was prior to 1960** if you delete the trust funds from the government's budget. **In 2002 the deficit was over one trillion dollars, in 2003 it was almost one and half trillion.**

The fourth key: Uncle Sam doesn't fund four of the major trusts—Social Security, federal employee retirement, military retirement, and railroad retirement. Benefits of these trusts are paid on a hand-to-mouth basis. Collected funds are needed and used to pay benefits before they can earn enough interest to make the trust self sustaining—it's called a Ponzi scheme. This practice is illegal in private business. Currently it takes approximately three people paying into Social Security to provide benefits to one person—it should be one-to-one. The government cannot properly fund the trusts with borrowed money because the interest alone would consume the government's total yearly receipts. This modus operadi is like a chain letter; it eventually runs out of enough people to feed it—Social Security predicts this will be case around 2030.

The fifth key: In four trusts there are two types of unfunded liabilities—an astronomical $17 trillion as of 2001 and now growing about $1 trillion a year. There are: 1) the liability to continue to pay benefits to present recipients until they are deceased, and 2) the liability to pay benefits in the future to persons who have been paying into the trusts, but because of age, etc. are not qualified as of now to receive benefits.

In short, the government's accounting practices are deceitful. They hide the true government's fiscal condition from the public. The President and Congress should be held accountable.

TABLE OF CONTENTS

CHAPTERS

APPENDICES

FIGURES

TABLES

THE WAY THINGS ARE

The blame for the economic collapse of our country ultimately will fall at all of our feet. But, to a large extent, it has its roots in a conspiracy of silence among presidential candidates and most members of Congress. That silence is due to basic fear. . . fear that if they confront the problem and honestly point out its severity to the American people, they may lose votes and elections.

— ***Senator Warren Rudman**, R-New Hampshire*
Comments on the Debt and Deficits in the
Foreword in Bankruptcy 1995

The U.S. decline began in 1970. It was hardly noticeable. As the 70s progressed, the public debt began to rise. It was accompanied by increased interest expense which rose faster than government receipts. Although contributory, the growing interest expense as a percentage of receipts was not the crux of the problem.

In 1978, William E. Simon, Secretary of the U.S. Treasury 1974-1977, wrote a book, *A Time for Truth*, warning the American people of the suicidal fiscal course our country was following. Storm warnings from other notable, knowledgeable, and highly respected persons were not far behind, but Congress and the public had deaf ears.

Figure 1-1 is a copy of the cover page of the September, 2001 issue of the *Monthly Treasury Statement of Receipts and Outlays*. This issue is for fiscal year 2001, which ends on September 30 2001. From this and statements for other years, the fiscal conditions of the government can be deduced.

In the 1960s, interest expense hovered around 11 percent; a number you, a family, a corporation, or the federal government could live with. But the rising interest expense, which began in 1970, reached 57.9 percent in 1991. Fed watchers thought the Armageddon was at hand. Fortunately, the subsequent boom times with larger government receipts rescued the U.S., reducing this frightening number to 34.4 percent by 2001. Chapter 5 discusses the Public Debt in some detail and data from 1960 through 2001 is provided. It's appalling to think that even now, one-third of the government's income goes to pay just interest expense. Even so, this shocking interest expense is not the crux of the problem.

As interest expense was growing, so were the trust funds. The trust funds are shown in Figure 1-2 which is a copy of Table 8 from the September 30th, 2001 issue of the *Monthly Treasury Statement of Receipts and Outlays.* By 1983, the trust funds had grown until they were gobbling up 50 percent of all government's receipts. The tail began wagging the dog. In 1993, trust funds receipts as a percent of all receipts had reached nearly 54 percent. Meanwhile trust fund outlays as a percent of all outlays were rising and by 2001, over 46 cents of every dollar the government spent was coming from the trust funds. The Department of Defense is often thought of as being a big spender, but compared to the trust funds it is small potatoes. When reference is made to all outlays by the government, it includes, in addition to the trust funds, the spending of the Office of the President, the Legislative Branch, the Judicial Branch, 14 departments which includes Defense, and some 60 independent, programs, foundations, commissions, corporation, boards, services, agencies, and what have you. Chapter 3 provides more details on the trust funds along with their receipts and outlays beginning in 1960 through 2003. As surprising as the preceding distribution of government outlays may be, this in itself is not the crux of the problem.

The crux of the decline and approaching fall of the U.S. will be attributable to the government's modus-operandi of four of the trust funds. The specific trust funds are currently listed in Figure 1-2 as 1) Federal Old Age and Survivors Insurance (Social Security), 2) Federal Employees Retirement, 3) Military Retirement, and 4) Railroad Retirement.

Final Monthly Treasury Statement

of Receipts and Outlays
of the United States Government

For Fiscal Year 2003 Through September 30, 2003, and Other Periods

Highlight

This issue includes the final budget results and details a deficit of $374.2 billion for Fiscal Year 2003.

Military active duty pay, veterans' benefits, supplemental security income payments, and Medicare payments to health maintenance organizations for September 1, 2003 were accelerated to August 29, 2003.

RECEIPTS, OUTLAYS, AND SURPLUS/DEFICIT THROUGH SEPTEMBER 2003

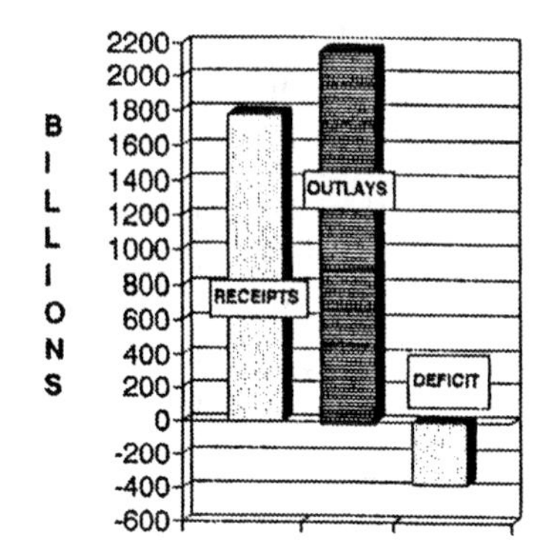

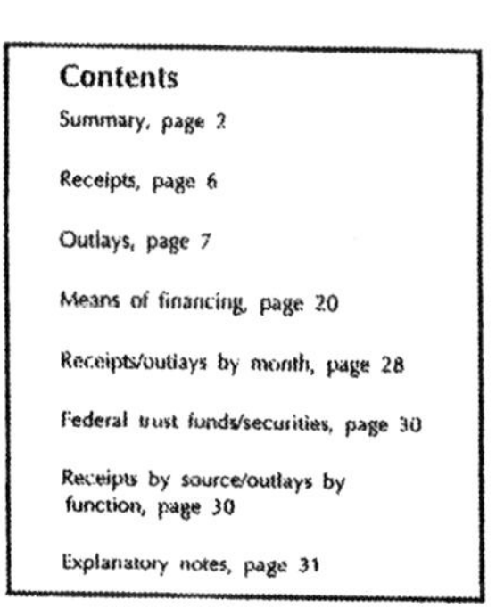
Contents

Compiled and Published by

Department of the Treasury
Financial Management Service

Figure 1-1

Table 8. Trust Fund Impact on Budget Results and Investment Holdings as of Septmeber 30, 2001

[$ millions]

Classification	This Month			Fiscal Year to Date			Securities held as Investments Current Fiscal Year		
							Beginning of		Close of This Month
	Receipts	Outlays	Excess	Receipts	Outlays	Excess	This Year	This Month	
Trust receipts, outlays, and investments held:									
Airport and airway	939	959	−21	10,044	9,524	519	13,097	14,396	13,660
Black lung disability	93	607	−514	524	1,013	−489	......	......	......
Federal disability insurance	6,196	−35	6,231	82,980	60,831	22,149	113,707	132,154	135,842
Federal employees life and health	......	−62	62	......	−1,978	1,978	28,362	30,360	30,341
Federal employees retirement	26,897	4,018	22,878	79,146	47,980	31,166	523,200	531,481	554,346
Federal hospital insurance	14,794	12,831	1,963	172,189	142,901	29,288	168,859	194,961	197,137
Federal old-age and survivors insurance	36,430	7,819	28,611	513,871	373,037	140,834	893,519	1,009,408	1,034,114
Federal supplementary medical insurance	337	7,000	−6,662	95,336	99,452	−4,116	45,075	48,917	41,978
Hazardous substance superfund	13	148	−135	940	1,242	−302	4,126	3,692	3,630
Highways	4,539	3,172	1,366	31,633	34,827	−3,194	31,023	26,155	24,115
Military advances	1,228	956	272	10,229	10,171	58	......	......	......
Military retirement	818	2,869	−2,050	40,826	34,096	6,730	149,348	159,119	156,978
Railroad retirement	491	697	−206	10,229	8,574	1,655	24,823	26,871	26,865
Unemployment	404	2,710	−2,307	33,993	31,623	2,370	86,399	91,107	88,638
Veterans life insurance	16	130	−114	1,055	1,186	−131	13,587	13,569	13,462
All other trust	361	464	−104	4,755	9,234	−4,480	14,088	14,369	14,341
Total trust fund receipts and outlays and investments held from Table 6-D	**93,554**	**44,284**	**49,271**	**1,087,748**	**863,712**	**224,036**	**2,109,212**	**2,296,560**	**2,335,447**
Less: Interfund transactions	31,364	31,364	......	349,940	349,940	......			
Trust fund receipts and outlays on the basis of Tables 4 & 5	62,190	12,919	49,271	737,809	513,773	224,036			
Total Federal fund receipts and outlays	**96,899**	**110,780**	**−13,881**	**1,253,518**	**1,350,390**	**−96,871**			
Less: Interfund transactions	594	594	......	1,124	1,124	......			
Federal fund receipts and outlays on the basis of Table 4 & 5	96,305	110,185	−13,881	1,252,394	1,349,266	−96,871			
Net budget receipts & outlays	**158,495**	**123,105**	**35,390**	**1,990,203**	**1,863,039**	**127,165**			

Note: Interfund receipts and outlays are transactions between Federal funds and trust funds such as Federal payments and contributions, and interest and profits on investments in Federal securities. They have no net effect on overall budget receipts and outlays since the receipts side of such transactions is offset against bugdet outlays. In this table, Interfund receipts are shown as an adjustment to arrive at total receipts and outlays of trust funds respectively.

Figure 1-2

The government's modus-operandi of three of these trusts is to use essentially current receipts to pay benefits. Government pabulum makes this sound innocuous, but nothing could be farther from the truth. It's a catch-22. Receipts are only in the trusts a fraction of the time necessary to garner the earnings needed to pay the specified benefits. Consequently to keep these trusts viable, it's necessary to have more persons paying into the trusts than persons receiving benefits. Like a dog chasing its tail, the trusts never catch up. This mode of operation is inherently flawed. It has the same failing as "chain letters"; you eventually run out of new victims. Currently, Social Security projects that benefits will exceed receipts in 2025 and by 2038 it will have completely exhausted it assets—a sign that the number of new persons paying into the trust is declining. As of 2001, the baby-boom generation, Americans born between 1946 and 1964, paying into the trust has temporarily raised the assets in the fund to its highest level in years. Even so, as of the close of fiscal year 2001 Social Security had only sufficient funds to pay benefits for 3 years of the 19 years needed, a mere shortfall of 16 years. Now it's the baby boomers' turn to receive their benefits . . . and we're not talking billions, the number is thousands of billions, trillions.

When private enterprise has used the procedure employed by the government to operate the four trust funds, it has been ruled illegal by the courts. It's called a Ponzi. Meanwhile the government marches to its own drummer while preaching "don't do as I do, do as I say"—and this is far from the only instance.

The unabridged Merriam-Webster defines Ponzi "(after Charles A. Ponzi, died 1949 American - Italian born swindler) as an investment swindle in which some early investors are paid off with the money put up by later ones in order to encourage more and bigger risks."

Since its inception in 1936, the number of covered workers in Social Security has steadily risen. To keep Social Security operable, each year the number of persons paying into the trust must exceed the number of persons receiving benefits by a factor of about three. Looking forward into the future, Social Security estimates this ratio will drop below three to one; that spells trouble with a capital T; that's when the baby-boomer start receiving benefits; that's why the estimated assets in the fund drop to zero in 2037; that's why the government's idea to let workers invest some of their Social Security taxes elsewhere simply won't fly. When members of the White House and Congress make this suggestion, it is a clear indication that they don't understand the problem or they're dishing out pabulum. Even with a ratio of three to one in the past, it has been necessary to raise taxes 20 times to keep Social Security viable. This upped the taxes collected on the earnings of individuals from 2% to 15.3%, a factor of over 7. A historical schedule of Social Security taxes and other data will be found in Appendix D.

The referral to Social Security as a Ponzi is not new. For a number of years, numerous references have been made to Social Security as a Ponzi. What's been missing are numbers—numbers which reveal the liability of current benefits—numbers which reveal the liability of future benefits—numbers which reveal the trend of these liabilities. And most important of all, what these numbers mean.

A quote from the Popular Lectures and Addresses of Lord William Thompson Kelvin in 1891-1894 illuminates the difference from yesterdays to today.

> "When you can measure what you are speaking about and express it in numbers you know something about it, but when you cannot measure it, when you cannot express it in numbers, your knowledge is a meager and unsatisfactory kind. It may be the beginning of knowledge but you have scarcely in your thoughts advanced to a stage of science."

And what are the numbers. On any given date, such as the end of fiscal year 2000, without receiving new funds Social Security should have sufficient assets to:

1) Pay benefits to all current beneficiaries until they are deceased and
2) Reimburse persons, who as yet are not qualified to receive benefits, what they have paid into the trust plus reasonable earning.

As indicated previously, at the close of fiscal year 2001 Social Security had only sufficient funds to pay benefits for 3 years, whereas benefits for 19 years are needed, a shortfall of $2.8 trillion—and this was the good news. There were participants who had been paying into the trust as long as 45 years, assuming age 20 as beginning work and 65 as retiring, and as yet were not old enough to qualify for benefits. Applying reasonable earnings to their contributions, they had a whopping equity of $11.6 trillion in the fund, making the total Social Security liability $14.4 trillion. To put this in perspective, this shortfall in Social Security alone is more than double the public debt of some $6 trillion. Keep in mind it takes 1,000 billion to equal a trillion. Thus any recent illusory surpluses of billions that Washington talks about wouldn't put a dent in these numbers. Then, there are the other three trust funds.

Funding current benefits with essentially current receipts is insidious. The undisclosed liabilities grow at an insatiable rate. In 1981 Social Security liabilities to pay just future benefits (the larger portion which excludes current benefits) was only $1.6 trillion, by 1989 it was $4.0 trillion, and in 2001 it was $11.6 trillion.

Returning to the present, for fiscal year 2001 when all of the unfunded liabilities are added to the public debt of about $6 trillion, the total is some $21 trillion. Just the interest on this overwhelming number is a staggering amount, about $1.2 trillion. In the year 2001 the total receipts from individual and corporate income taxes, and all other revenues, excluding the trust funds, were only $0.991 trillion. Meanwhile, the public believes Social Security is OK until around 2030.

Future projections are astronomical. Using data provided by Social Security and projecting liabilities as of 2038, brings up the incredible number of $147.2 trillion, see Appendix C. There seems to be little doubt that we have passed the point of no return, which is, the government being able to make good on its commitments. When you can't pay the interest on your liabilities, you're bankrupt. It is simply a matter of time. As sure as the earth rotates, these inherently faulty Ponzi trust operations will have to be changed and the liabilities of these trust funds restructured—referred to as a Chapter 11 in private industry. There is a limit on the number of new participants and/or how much higher taxes or contributions can be raised. But, as our government has demonstrated, Ponzis can be kept operating for a long time, years. However, it should be noted, the quicker they fail, the lesser the damage, so the sooner the better.

The Appendices provide the data and information on how relatively simple calculations can reveal the liability numbers. The actual liability calculations for the four trust funds and projections for Social Security are included in separate appendices. The underlying data has been obtained from the U.S. Department of the Treasury and the individual trust funds.

Clearly our people in Washington don't see the danger in these Ponzi operations; they're making plans to reduce taxes. President Bush wasn't even aware that Social Security was a government trust fund. But then, this must have been the kind of thinking that prevailed in other countries when their decline, from number one in the world, got under way—to name the more recent, Spain and England.

Several questions arise at this point: 1) when this information becomes public, what's Washington reaction likely to be, and 2) when will the condition of the U.S. be generally recognized. These questions and others will be addressed in subsequent chapters.

PRIOR STORM WARNINGS

One thing is certain. At some point global investors will lose confidence in our (U.S.) easy dollars and debt-financed prosperity, and then the chickens will come home to roost.

— David A. Stockman[1]
Director of the Office of Management and Budget during the Reagan Administration

Although we have seen some nasty recessions and sharp setbacks in the stock market over the last three decades, on balance the good times have rolled and affluence has become a way of life for many U.S. citizens. Advances in technology have apparently made yesterday's ways and machines obsolete, creating unending new products and markets. It would seem that depressions have become obsolete and historic; that doom and gloom are mere figments of unsound thinking. And in fact, the fiscal health of our country seems to have improved significantly since 1991 when 57.9% of usable receipts were required to pay interest expense on the public debt; when it appeared that the doom and gloomers' dire predictions were in the wings and about to make their debut.

The down turn in the economy of many foreign countries in the nineties had little effect on the U.S. We appeared to be immune to their malaise. With the announcement that the U.S. had a fiscal surplus in 1998, 1999 and 2000, the general perception was our government's finances are in good shape.

In the past there have been people reminding the public of our government's fiscal improprieties; certainly for at least the last 50 years; perhaps ever since our government was first formed in 1776. However, since World War II they had little basis in fact until the mid 1970s. They were simply cries of "wolf." But as the 1970s progressed, the annual government deficits and the total public debt began to rise ominously. Regardless of which political party was in power, our nation's fiscal ills continued to go from bad to worse, not to mention our other problems such as health care, social security, handling of foreign situations, etc. A review of some of the dire and past forecasts, which didn't materialize but had a basis in fact, is in order to gain a perspective of where we are today and where we are headed.

Perhaps the first realistic and verifiable public warning of future fiscal problems came in 1978 from Mr. William E. Simon who was Deputy Secretary of the Treasury from 1973-1974 and Secretary of the Treasury from 1974-1977. In Mr. Simon's book, *A Time for Truth*[2], he warned that our profligate ways were an invitation to disaster. Also in his book, Mr. Simon reproduces the following relevant portion of a statement he made to the Subcommittee on Democratic Research Organization of the House of Representatives. "We all know that neither man nor business nor government can spend more than is taken in for very long. If it continues, the result must be bankruptcy."[3] In his words, "We have continued to proliferate both the recorded and unavowed debt at rates substantially greater than the growth of our ability to pay. The visible or acknowledged debt alone has grown at an annual rate of 14 percent in the past decade, while the GNP (Gross National Product) has grown at a far lower rate. Clearly this system cannot go on indefinitely."[4] His book was lauded by two Nobel Laureates, Milton Friedman and F. S. Hayek, but unfortunately his words had negligible impact, if any, on Washington or the public.

Our government's annual deficits and the public debt continued to mount as we moved into the 1980s. The debt of $996.6 billion accumulated by our government in the 6 years beginning in 1980 through 1985, exceeded the total debt of $826.5 billion incurred in the 204 years prior to 1980, that is, between 1979 and when our country first came into being in 1776.

A large portion of the frightening increase in the debt during the early 80s can be attributed to President Reagan's emphasis on military spending, which included Star Wars, an ill-fated adventure. Contrarily, President Reagan in 1982, with seemingly good intentions, tried to implement his campaign pledge to rein in government spending, reduce the annual deficits and the public debt. In his words, repeated many times during his political campaigns, "he was going to get the government off the backs of the American people." At a press conference on February 18, 1982 the President announced the formation of the President's Private Sector on Cost Control. President Reagan pledged: "This is not going to be just another blue-ribbon, ornamental panel. We mean business, and we intend to get results." At that time, Reagan named J. Peter Grace, Chairman of W. R. Grace & Co., to direct this effort, which later became known as the Grace Commission. The Commission was formally launched on June 30, 1982, when the president issued Executive Order 12369.

The Grace Commission was composed of 36 task-force teams, chaired by 161 highly respected and well-known individuals from corporate, academic, and labor positions. The total staff consisted of over 2,000 volunteers. During some 18 months of rigorous evaluation of the federal government, the Commission compiled 2,478 separate and carefully projected recommendations[5]. It was estimated the implementation of these recommendations could save the government about $141 billion each year; at that time cutting annual government outlays by approximately 18%. An overview of the Grace Commission will be found in Appendix H. The Commission's expertise is apparent from the list of prominent and highly respected people who chaired or co-chaired specific task force teams responsible for seeking ways to reduce the cost of our government.

When the Commission had completed its task and released its report, there was little argument that, on the whole, the recommendations were good, viable, and could indeed save the government billions of dollars. However, when it came to implementation that was another story. One would be hard put to point to a single Grace Commission recommendation that ever got partially or fully implemented. Maybe it was a case of not being conceived, supported and paid for by Congress; it was the President's idea and private industry picked up the entire tab which was about $75 million. The bottom line was the debt and the annual deficits kept growing at an unprecedented rate. To Mr. Grace's credit, he was undaunted. He became a Founder and Co-chairman of the "Citizens Against Government Waste,"[6] one of the leading public advocates of fiscal reform of our government.

Another prominent business leader, Lee Iacocca, spoke out in 1986. "I agree with Hadady that the interest bill on the federal debt ought to put Uncle Sam in the intensive care unit. I'm getting a little nervous that so many people still can't see how sick the old gentleman (Uncle Sam) is."[7] Mr. Iacocca also expressed his concerns in considerable detail in a feature column that appeared irregularly in major newspapers around the country for about a year. Later, in his book, *Talking Straight*[8], published in 1988, he spelled out in no uncertain terms his deep concern about the fiscal course our government was pursuing.

In the second half of the 1980s, the public heard rumblings in Congress of our exorbitant spending habits. Some members, specifically Senator Phil Gramm, Senator Warren B. Rudman, and Senator Ernest F. Hollings, sponsored the Balanced Budget and Emergency Deficit Control Act, PL 99-177, which was enacted in December, 1985. This Act, referred to by the names of its principal sponsors, Gramm-Rudman-Hollings, committed both the President and Congress to a fixed schedule of progress toward reducing the deficit. Twelve members of Congress led by Democrat Representative Mike Synar from Oklahoma and the National Treasury Employees Union challenged the Act in court. Seven months later, the Supreme Court in ruling 478 U.S. 714, on July 7, 1986 declared the Act unconstitutional[9]. In a 7-2 decision, the court said the law's mechanism for automatic spending cuts was unconstitutional; a violation of the principle of separation of powers in that it vested executive-branch authority in a legislative-branch officer, the Comptroller General. The dissenting Justices were Byron R. White and Harry A. Blackmun

Deeply stirred by the Supreme Court's ruling declaring his deficit reduction law unconstitutional and his lack of success in reducing government spending, Senator Rudman decided to retire and not run for re-election in 1992. He considered the deficit "immoral." Retiring Senator Rudman had this to say about our debt and deficits in the foreword of *Bankruptcy 1995*[10], "The blame for the economic collapse of our country ultimately will fall at all of our feet. But, to a large extent, it has its roots in a conspiracy of silence among presidential candidates and most members of Congress. That silence is due to basic fear. . . fear that if they confront the problem and honestly point out its severity to the American people, they may lose votes and elections."

In 1991, the prestigious Conference Board[11] surveyed 5000 households to determine which public issues were of greater concern. The Federal Budget Deficit was third in the survey, led only by the cost of Medical Care and Drug Abuse.

In the first half of 1992, a proposed Amendment to the U.S. Constitution was submitted to Congress, S.J. Res.41. The Amendment required the federal government to have a balanced budget by the year 2001. Further, expenditures could not exceed receipts in any year unless the U.S. was at war or three-fifths of the membership of both houses of Congress voted to override the requirement in any year. The amendment was defeated in the House on June 11, 1992, by a vote of 63-37.

In September 1992, Senator Warren Rudman, Paul Tsongas, former Democratic presidential candidate, and Peter Peterson, Commerce Secretary during the Richard M. Nixon Administration, formed the Concord Coalition[12]. The object of the organization was to educate the public about the deficit. One of their means for doing so was a quarterly publication, the "Concord Courier."

Still the annual deficits and the debt rose unabated. By September 30, 1993 the public debt, according to Treasury Statements, was some $17,006 for each man, woman, and child in the U.S.[13]

Even gifts to reduce the Public Debt were made. In 1992 these gifts amounted to $ 4,547,927.14; in 1993, $1,726,763.07[14]. Although well intended, these gifts were hardly sufficient to get recorded since the numbers in the "Monthly Statement of the Public Debt" are reported in millions.

By the close of fiscal 1993, the gross interest expense on the debt had exceeded military expenditures by the Department of Defense and had become the second largest item in the budget, exceeded only marginally by Social Security. However, the deficit was so large that if interest payments to non-government lenders of $199 billion could have been skipped, we would still have run a deficit of $56 billion. But U.S. citizens had become inured to the cry of "wolf."

Our government's fiscal improprieties have been well known for the past twenty years, but somehow our nation keeps muddling along and Congress, for the most part, has deaf ears.

One of the most fiscally knowledgeable persons in our country, the head of our nation's bank, Federal Reserve Chairman, Alan Greenspan, has repeatedly expressed his deep concern to Congressional committees and others about our spendthrift ways. Appearing before the Committee on Finance of the U.S. Senate in 1993 he said, "Let me conclude by reiterating my central message. The deficit is a malignant force in our economy. How the deficit is reduced is very important; that it be done is crucial. Allowing it to fester would court a dangerous erosion of our economic strength and a potentially significant deterioration in our real standard of living."[15].

Greenspan's predecessor, Paul A. Volcker, was likewise as concerned. In his testimony before a Congressional Committee on February 7, 1984, he stated, "Our common sense tells us that enormous and potentially rising budget deficits, and the high and rising deficits in our trade accounts, are wrong—they can not be indefinitely prolonged." Later in his testimony, he said, "It is already late. The stakes are large, markets have a mind of their own; they have never waited on the convenience of kings or Congressmen—or elections." Two years later in testifying before the House Ways and Means subcommittee on September 24, 1986, Mr. Volker stated that the United States has been living in a "false paradise" in which we have grown dependent on heavy flows of foreign capital to keep our standard of living rising.

In 1994, Senator Paul Simon (D-Illinois) and Senator Larry Craig (R-Idaho), co-authored an article "Our Economic Security in the Balance,"[16] citing the danger of our government's addiction to debt. At least there are a few members of Congress who appreciate the danger of our profligate ways; the problem has been and is there just aren't enough of them.

During the 80s and 90s, a number of non-profit organizations were formed; their goal to reduce government waste and balance our national budget via an informed populace. These organizations[17] included "Citizens Against Government Waste," "National Tax Payers Union," "Heritage Foundation," the "Concord Group," and others. Government waste has been and continues to be a popular subject. Although it may seem difficult to believe, supporters to reduce waste are also found within our government's civil service. The Congressional Budget Office, Office of Management and Budget, and the HHS Inspector General have all provided documentation on how government expenses could be significantly reduced. Their efforts, like other attempts, have been futile in slowing the growth of the government's outlays. Everyone, except the congressional body, is seemingly convinced that our government is squandering our money.

There are also a number of editors of nationally recognized financial market letters who have been saying our government is headed for bankruptcy or is already bankrupt. When you're talking about the U.S. dollar going down the tubes, the de facto currency of the world, bankruptcy is highly improbable. Keep in mind the U.S. issues and in large measure controls their currency. The U.S. can't just close the doors, sell off its assets, and pay off the creditors with whatever is left in the till. However, a possible big devaluation of the dollar is another story.

Serious warnings about our grasshopper ways have been flying for the last 20 years; Congress and the public have failed to do anything. And why should anyone, when the good times continue to roll. Clearly there has been a lot of smoke but no signs of a fire. Meanwhile, the majority of our girls and boys on Capitol Hill continue to play their political games while keeping an eagle eye on their holy of holies, the polls, to keep them informed on what to do to get re-elected. It's no wonder that most people have become inured to any dire predictions that come their way.

Perhaps it is what these seers have **not** said or are **not** saying that has inured us. For the most part, seers have repeatedly told us that it will be our children who suffer the consequences of our spendthrift ways, a problem off in the hazy future and one that will not confront us directly. Certainly the recent boom times of the 90s have pushed a confrontation a number of years into the future.

Most doomsayers state that our government is or will become bankrupt, but few say how or when or its effects. Most identify areas where our government is wasting money, where large savings can be made, and how to get us back on the right track . . . as if this information were being revealed for the first time. This isn't and hasn't been the problem for years. Even the most remote recluse is well aware of what's needed. Government outlays have to be cut and/or taxes raised to generate large surpluses to reduce the public debt and the liabilities of the trusts. The problem is implementation.

The great communicator, President Reagan, bolstered with the data from his Grace Commission, wasn't able to get the government off the backs of the American people, so it seems improbable that anyone else could get the job done. On the other hand, although it seems unlikely, maybe President Reagan's words were only political harangue without serious intent. Regardless of who or how it is done, it ain't going to easy. In short, the likelihood of it getting done is somewhere between slim and none.

Since 1978 some of the better financial minds in the country have been predicting that the U.S. is headed for deep fiscal trouble. Some had the temerity to predict our problems would surface in years already behind us. Others were indefinite, which isn't much help, but it seems unlikely that any of them were peering as far ahead as the current date.

In retrospect, the public debt was and is still very serious. Its presence, known to all, attracted and has held the focus of our populace and the government officials. Meanwhile, the liabilities of the seemingly innocuous trust funds have been growing cancer-like. These liabilities are now so large the public debt is small potatoes. A currency devaluation, monetizing the debt or whatever is incapable of resolving the problems. Resolution can only come about by restructuring these liabilities; liabilities which involve funds needed for aid and retirement of two-thirds of our elderly. The result will be catastrophic because such a large portion of our population is involved. The doomsayers will have turned out to be right, but for the wrong reasons. The days of being the greatest nation on earth will come to an end. We'll be replaced. History has shown there is always a nation waiting in the wings.

References:

1) Stockman, David A. *The Triumph of Politics - Why the Reagan Revolution Failed.* New York, NY: Harper & Row, Publishers, 1986: p.377

2) Simon, William E. *A Time for Truth,* New York: McGraw-Hill Book Company, 1978

3) Simon, William E. *A Time for Truth,* New York: McGraw-Hill Book Company, 1978: p. 11

4) Simon, William E. *A Time for Truth,* New York: McGraw-Hill Book Company, 1978: p. 100-101

5) Kennedy, Jr., William R. and Lee, Robert W., *A Taxpayer Survey of the Grace Commission Report,* Ottawa, IL, Green Hill Publishers, 1984.

6) Citizens Against Government Waste, 1301 Connecticut Avenue, NW - Suite 400, Washington, DC 20036, Phone (202) 467-5300.

7) Hadady, R. Earl. *How Sick is Uncle Sam.* Pasadena, CA: Key Books Press, 1986.

8) Iacocca, Lee. *Talking Straight.* New York, NY: Bantam Books, 1988

9) "Congress and the Nation." Congressional Quarterly. Washington, DC: U.S. Government Printing Office, Volume VII, 1985-1988, p.834

10) Figgie, Jr., Harry E. with Swanson, Ph.D., Gerald J. *Bankruptcy 1995, The Coming Collapse of America and How to Stop It.* Boston: Little, Brown and Company, 1992, Foreword, p. xiii

11) The Conference Board, 845 Third Avenue, New York, NY 10022. Phone (212) 759-0900. The Conference Board is an independent, not-for-profit research institution engaged in objective studies of management and economics.

12) The Concord Coalition, 1025 Vermont Avenue, NW - Suite 810, Washington, DC 20005. Phone (202) 737-1077

13) Based on the Public Debt of $4,411,489 million as reported in the September 30, 1993 issue of the U.S. Treasury document, "Monthly Statement of the Public Debt of the United States," and an estimated population of 259.4 million.

14) "Monthly Statement of the Public Debt of the United States," September 30, 1993, Table V, p.25, Superintendent of Documents, U.S. Government Printing Office, Washington, DC 20402

15) Statement by Alan Greenspan, Chairman, Board of Governors of the Federal Reserve System before the Committee on Finance, U.S. Senate, March 24, 1993, Statements to Congress, p. 473

16) "Government Waste Watch," Winter 1994, p.6. Newsletter of "Citizens Against Government Waste," 1301 Connecticut Avenue, NW - Suite 400, Washington, DC 20036, Phone (202) 467-5300.

17) Citizens Against Government Waste, 1301 Connecticut Avenue, NW - Suite 400, Washington, DC 20036, Phone (202) 467-5300. National Taxpayers Union, 325 Pennsylvania Avenue, SE, Washington, DC 20003, Phone (202) 543-1300. Heritage Foundation, 214 Massachusetts Avenue, NE, Washington, DC 20002, Phone (202) 546-4400.

TRUST FUNDS

Trust is a loose word in the government's vocabulary
— R. Earl Hadady

In the past, Congress established various trusts to fund special programs, such as Social Security, federal employee retirement, unemployment, hazardous substance disposal, veterans life insurance, etc. Likewise, in most cases special arrangements were made to fund these trusts as opposed to using general government revenues.

Trust funds are just what the name implies, funds set aside for a specific use and no other. Monies in these trusts should be considered sacrosanct.

Some of the trusts are funded by excise taxes collected by the Internal Revenue Service (IRS), a division of the Treasury Department. Speaking of this all-time favorite government service, the IRS, they got a bit of comeuppance in a 1992 audit by the General Accounting Office (GAO), the investigative arm of Congress. They found that the money the Treasury Department forwards to the trust funds is, to a large extent, guess work. What a sloppy way to do business! You can imagine what the IRS would do to you if you pulled this kind of shenanigans. Charles A. Bowsher, the GAO's comptroller general had this to say about the IRS. "Internal controls were not properly designed and implemented to effectively safeguard assets or provide a reasonable basis for determining material compliance with laws governing the use of budget authority and other relevant laws and regulations."[1]

Mr. Bowsher went on to say that the information the IRS lacks includes such basic data as how much money it has collected from different excise taxes. Those taxes are earmarked for trust funds that are dedicated to specific purposes, such as building highways or cleaning up toxic wastes. Lacking IRS information on how much has been collected, the GAO said, the money the Treasury Department forwards to the trust funds is based on IRS assessments, not actual collections. The result is some fiscal inaccuracies in data for the trust funds which represents the largest portion of the federal budget.

Figure 3.1 is a copy of Table 8 from the *Monthly Treasury Statement of Receipts and Outlays of the United States Government* which covers the Trust Funds.

Table 3.1 provides a listing of the trusts in the order of descending holdings of federal securities, the only type of security allowed by law.

Tables 3.2 and 3.3 show receipts and outlays of the trust funds versus total government receipts and outlays. **In 2001, 47 cents of every dollar the government collected went into the trust funds; 46 cents out of every dollar spent by the government was spent by the trust funds.** Astounding isn't it. When you consider the rest of the government includes the Legislative Branch, the Judicial Branch, the Executive Branch with its 13 departments which includes Defense. Then you must add some 50 independent establishments and government corporations. The proverbial big spender, the Department of Defense, accounted for only 15.6% of the total outlays in 2001; Secretary Rumsfeld may have mixed feelings about learning he is running only a nickel and dime operation.

The following is a brief description of each of the major trust funds.

Airport Trust

Purpose: Meet obligations for airport planning and development and noise compatibility planning and programs; facilities and equipment; research, engineering and development; and a portion of operations.

Funded by: Passenger ticket tax and certain other taxes paid by airports and airway users.

Black Lung Disability Trust

Purpose: Pays compensation, medical, and survivor benefits to eligible miners and their survivors, where mine employment terminated prior to 1970 or where no mine operator can be assigned liability.

Funded by: Excise tax on mined coal tonnage

Table 8. Trust Fund Impact on Budget Results and Investment Holdings as of Septmeber 30, 2001

[$ millions]

Classification	This Month			Fiscal Year to Date			Securities held as Investments Current Fiscal Year		
							Beginning of		Close of This Month
	Receipts	Outlays	Excess	Receipts	Outlays	Excess	This Year	This Month	
Trust receipts, outlays, and investments held:									
Airport and airway	939	959	-21	10,044	9,524	519	13,097	14,396	13,660
Black lung disability	93	607	-514	524	1,013	-489	......	......	......
Federal disability insurance	6,196	-35	6,231	82,980	60,831	22,149	113,707	132,154	135,842
Federal employees life and health	......	-62	62	......	-1,978	1,978	28,362	30,360	30,341
Federal employees retirement	26,897	4,018	22,878	79,146	47,980	31,166	523,200	531,481	554,346
Federal hospital insurance	14,794	12,831	1,963	172,189	142,901	29,288	168,859	194,961	197,137
Federal old-age and survivors insurance	36,430	7,819	28,611	513,871	373,037	140,834	893,519	1,009,408	1,034,114
Federal supplementary medical insurance	337	7,000	-6,662	95,336	99,452	-4,116	45,075	48,917	41,978
Hazardous substance superfund	13	148	-135	940	1,242	-302	4,126	3,692	3,630
Highways	4,539	3,172	1,366	31,633	34,827	-3,194	31,023	26,155	24,115
Military advances	1,228	956	272	10,229	10,171	58	......	......	......
Military retirement	818	2,869	-2,050	40,826	34,096	6,730	149,348	159,119	156,978
Railroad retirement	491	697	-206	10,229	8,574	1,655	24,823	26,871	26,865
Unemployment	404	2,710	-2,307	33,993	31,623	2,370	86,399	91,107	88,638
Veterans life insurance	16	130	-114	1,055	1,186	-131	13,587	13,569	13,462
All other trust	361	464	-104	4,755	9,234	-4,480	14,088	14,369	14,341
Total trust fund receipts and outlays and investments held from Table 6-D	**93,554**	**44,284**	**49,271**	**1,087,748**	**863,712**	**224,036**	**2,109,212**	**2,296,560**	**2,335,447**
Less: Interfund transactions	31,364	31,364	......	349,940	349,940	......			
Trust fund receipts and outlays on the basis of Tables 4 & 5	62,190	12,919	49,271	737,809	513,773	224,036			
Total Federal fund receipts and outlays	**96,899**	**110,780**	**-13,881**	**1,253,518**	**1,350,390**	**-96,871**			
Less: Interfund transactions	594	594	......	1,124	1,124	......			
Federal fund receipts and outlays on the basis of Table 4 & 5	96,305	110,185	-13,881	1,252,394	1,349,266	-96,871			
Net budget receipts & outlays	**158,495**	**123,105**	**35,390**	**1,990,203**	**1,863,039**	**127,165**			

Note: Interfund receipts and outlays are transactions between Federal funds and trust funds such as Federal payments and contributions, and interest and profits on investments in Federal securities. They have no net effect on overall budget receipts and outlays since the receipts side of such transactions is offset against bugdet outlays. In this table, Interfund receipts are shown as an adjustment to arrive at total receipts and outlays of trust funds respectively.

Figure 3-1

Table 3.1

Trust Funds Assets

Listed in order of decreasing assets
as of the close of fiscal year 2001

	Trust Name	**Assets** $s millions (1)
1	Federal old-age and survivors insurance (2)	1,034,114
2	Federal employees retirement	554,346
3	Federal hospital insurance	197,137
4	Military retirement	156,978
5	Federal disability insurance	135,842
6	Unemployment	88,638
7	Federal supplementary medical insurance	41,978
8	Federal employees life and health	30,341
9	Railroad retirement	26,865
10	Highways	24,115
11	All other trusts	14,341
12	Airport and airway	13,660
13	Veterans life insurance	13,462
14	Hazardous substance superfund	3,630
15	Black lung disability	0
16	Military advances	0

(1) Data from Table 8, p. 30, Sept. 2001 issue of the *Monthly Treasury Statement of Receipts and Outlays*

(2) Social Security

Table 3.2

Receipts of Trust Funds versus All Others

Year	Receipts				
	Total	**Trusts**			
	Government	Receipts & Interest	Interest Only	Receipts Net	Net Receipts % of Total
Fiscal	$s m	$s m (1)	$s m(2)	$s m	Government
1982	617,766	269,266	15,991	253,275	41.0%
1983	600,563	317,670	16,952	300,718	50.1%
1984	666,457	335,364	20,333	315,031	47.3%
1985	734,057	394,468	26,047	368,421	50.2%
1986	769,091	420,273	27,873	392,400	51.0%
1987	854,143	461,820	35,015	426,805	50.0%
1988	908,166	510,861	41,822	469,039	51.6%
1989	990,789	557,709	51,861	505,848	51.1%
1990	1,031,308	590,754	62,312	528,442	51.2%
1991	1,054,260	631,188	70,649	560,539	53.2%
1992	1,091,692	663,468	77,838	585,630	53.6%
1993	**1,153,175**	**702,771**	**82,276**	**620,495**	**53.8%**
1994	1,257,187	726,179	85,698	640,481	50.9%
1995	1,350,576	763,281	93,176	670,105	49.6%
1996	1,452,763	816,233	98,029	718,204	49.4%
1997	1,578,977	859,468	104,992	754,476	47.8%
1998	1,721,421	909,333	113,838	795,495	46.2%
1999	1,827,285	979,367	118,634	860,733	47.1%
2000	2,025,038	1,033,531	128,911	904,620	44.7%
2001	1,990,203	1,087,748	143,935	943,813	47.4%
2002	1,853,288	1,131,034	153,195	977,839	52.8%
2003	**1,782,317**	**1,167,823**	**156,111**	**1,011,712**	**56.8%**

Notes: The *Monthly Treasury Statement* provides only 1) Receipts

Table 3.3

Outlays of Trust Funds versus All Others

Year	Outlays		
	Government Total	Trusts Total	Trusts % of Government
Fiscal	$s m	$s m	
1982	745,706	263,810	35.4%
1983	808,327	294,613	36.4%
1984	851,781	304,799	35.8%
1985	946,316	340,928	36.0%
1986	990,231	420,273	42.4%
1987	1,003,804	388,404	38.7%
1988	1,063,318	413,440	38.9%
1989	1,144,020	434,040	37.9%
1990	1,251,776	474,830	37.9%
1991	1,322,561	519,024	39.2%
1992	1,381,895	567,545	41.1%
1993	1,408,122	601,209	42.7%
1994	1,460,557	630,877	43.2%
1995	1,514,389	664,521	43.9%
1996	1,560,094	701,429	45.0%
1997	1,601,595	735,745	45.9%
1998	1,651,383	753,523	45.6%
1999	1,704,545	772,000	45.3%
2000	1,788,045	804,807	45.0%
2001	**1,863,039**	**863,712**	**46.4%**
2002	2,011,808	929,545	46.2%
2003	2,156,536	989,585	45.9%

Federal Old-Age and Survivors Insurance Trust
(Social Security)

Purpose: Provide income to retired workers, their dependents, and survivors.

Funded by: Payroll deduction, employer contributions, and self-employed.

Federal Disability Insurance Trust

Purpose: Provides income to insured disabled workers, their dependents, and selected others including military reservists on inactive duty training, some agricultural workers, etc.

Funded by: Payroll taxes paid by workers, employers, and self-employed.

Federal Hospital Insurance Trust
(Medicare, Part A)

Purpose: Funds the costs of hospital and related care for most individuals age 65 or older and for disabled people.

Funded by: Non-optional deductions from Social Security payments.

Federal Supplementary Medical Insurance Trust
(Medicare, Part B)

Purpose: Affords protection against the costs of physician and certain other medical services.

Funded by: Optional deductions from payments into Social Security fund about 25% of the cost of this program. The remainder is funded from general government revenues.

Highway Trust

Purpose: Aid states in highway construction and maintenance.

Funded by: Motor fuel tax and certain other taxes paid by highway users.

Unemployment Trust

Purpose: Aid states in providing temporary income to qualified individuals who become unemployed.

Funded by: Payroll taxes

Railroad Retirement Trust

Purpose: Provide income to retired railroad workers

Funded by: Federal Disability Insurance Trust Fund, Social Security and payroll deduction.

Federal Employees Life & Health Trust

Purpose: Provide life insurance and health benefits to federal employees.

Funded by: Contributions from those participants enrolled in the government-sponsored plan and by government contributions.

Federal Employees Retirement Trust

Purpose: Provide income for retired federal employees

Funded by: Contributions from federal employees via payroll deduction plus government contributions.

Military Advances Trust

Purpose: An account for holding foreign government deposits for the purpose of buying U.S. military arms. Established by the 1976 Arms Export Control Act.

Funded by: Foreign governments. Funds must be deposited prior to arms shipment and may or may not earn interest, depending on the result of negotiations between the foreign government and the U.S. Defense Security Assistance Agency.

Military Retirement Trust

Purpose: Provide retirement income for military personnel.

Funded by: Funded entirely by the government.

Veterans Life Insurance Trust

Purpose: Provide veterans with low-cost life insurance.

Funded by: Payroll deductions. Since 1965, this program has provided low cost term insurance protection to members of the armed services through a group policy issued by the Prudential Life Insurance Company. Under the policy, the government agreed to pay the cost of the claims that were incurred as a result of the extra hazards of service. Between 1956 and 1965 no policies were issued except to those persons with a prior service connected disability. Prior to 1956, the government was the insurer.

Hazardous Substance Superfund Trust

Purpose: Cleaning up hazardous substance emergencies and uncontrolled waste sites.

Funded by: Cost shared by the Federal and State governments as well as industry. EPA will allocate funds from its appropriations to other Federal agencies to carry out the Act.

Various other minor trusts

References:

1) "GAO Finds Huge Gaps in IRS Records," Los Angeles Times, July 2, 1993, Section D, p.2

SURPLUS DELUSIONS

A billion here and a billion there, and soon you're talking about real money.

— ***Everett McKinley Dirksen (1896–1969)***
US Senator from Illinois

The old Washington "smoke and mirrors" trick was brought out of the closet by President Clinton when the Treasury Statement for the fiscal year 1998 became available. The President announced that we had a budget surplus, the first since 1969. He recommended using the surplus to bolster the Social Security trust fund, a suggestion that undoubtedly played well in Peoria.

Figure 4-1 is Table 3 on page 5 of the *Final Monthly Treasury Statement of Receipts and Outlays of the United States Government for Fiscal Years 1999.* This statement was selected because it includes data for both 1999 and 1998—see table headings Current Fiscal Year to Date and Comparable Prior Period. It was the 1998 data in this table that enabled the President to beat his chest and proclaim a surplus. The specific data are toward the top of the page and the bottom of the page. Near the top of the page, data is listed under Budget Receipts, Total Receipts, and (Off-budget). Close to the bottom of the page, data is listed in the last three rows as Surplus (+) or deficit (-), (On-budget) and (Off-budget). Note, the (On-budget) numbers at the bottom of the page were both negative for 1999 and 1998. It was the (Off-budget) numbers that enabled the president to declare a surplus.

And just what are the (Off-budget) items. In Figure 4-1 (Table 3 on page 5 of the Treasury statement) near the top of the page, the last item under Budget Receipts is (Off-budget) which lists receipts of $444,468m for 1999. The source of these receipts may be found by referring to Figure 4-2 (Table 4 on page 6 of the Treasury statement). Under the left hand heading, Classification, and part way down the

Table 3. Summary of Receipts and Outlays of the U.S. Government, September 1999 and Other Periods
[$ millions]

Classification	This Month	Current Fiscal Year to Date	Comparable Prior Period	Budget Estimates Full Fiscal Year[1]
Budget Receipts				
Individual income taxes	89,250	[2]879,480	828,587	886,657
Corporation income taxes	40,235	184,680	188,877	179,494
Social insurance and retirement receipts:				
Employment and general retirement (off-budget)	39,093	444,468	415,800	444,416
Employment and general retirement (on-budget)	15,701	136,411	124,215	132,567
Unemployment insurance	332	26,480	27,484	26,719
Other retirement	356	4,472	4,335	4,319
Excise taxes	7,167	70,399	57,669	70,655
Estate and gift taxes	2,294	27,782	24,076	28,441
Customs duties	1,727	18,336	18,297	17,994
Miscellaneous receipts	4,242	34,777	32,325	35,078
Total Receipts	**200,396**	**1,827,285**	**1,721,465**	**1,826,340**
(On-budget)	**161,304**	**1,382,817**	**1,305,666**	**1,381,924**
(Off-budget)	**39,093**	**444,468**	**415,800**	**444,416**
Budget Outlays				
Legislative Branch	201	2,621	2,600	2,853
Judicial Branch	317	3,793	3,463	3,912
Department of Agriculture	4,399	62,885	53,950	62,678
Department of Commerce	460	5,036	4,047	4,796
Department of Defense—Military	22,951	261,379	256,124	288,570
Department of Education	3,492	33,521	31,498	34,323
Department of Energy	1,551	16,079	14,444	15,544
Department of Health and Human Services	31,187	[3]359,700	350,571	371,260
Department of Housing and Urban Development	4,776	32,736	30,224	32,988
Department of the Interior	805	7,773	7,232	8,580
Department of Justice	1,661	16,318	16,169	18,648
Department of Labor	2,856	32,459	30,002	32,885
Department of State	497	6,464	6,373	7,034
Department of Transportation	4,519	41,819	39,467	41,913
Department of the Treasury:				
Interest on the Public Debt	19,786	353,511	363,824	354,651
Other	−2,106	[2,3]33,769	26,270	33,799
Department of Veterans Affairs	3,633	43,169	41,776	43,913
Corps of Engineers	490	4,186	3,833	4,209
Other Defense Civil Programs	2,662	32,008	31,216	32,311
Environmental Protection Agency	609	6,752	6,288	6,666
Executive Office of the President	21	416	236	387
Federal Emergency Management Agency	151	4,040	2,101	3,120
General Services Administration	−69	−46	1,095	361
International Assistance Program	760	10,059	8,980	10,497
National Aeronautics and Space Administration	1,261	13,664	14,206	14,043
National Science Foundation	299	3,285	3,188	3,259
Office of Personnel Management	4,363	47,515	46,307	48,268
Small Business Administration	249	[4]58	−78	−814
Social Security Administration	35,019	419,790	408,202	420,509
Other independent agencies	5,582	6,865	10,653	6,482
Allowances	......	......	......	632
Undistributed offsetting receipts:				
Interest	−1,052	−118,634	−113,839	−120,388
Other	−7,164	−40,446	−47,197	−40,387
Total outlays	**143,966**	**1,704,646**	**1,652,224**	**1,727,502**
(On-budget)	**108,846**	**1,383,767**	**1,335,622**	**1,406,683**
(Off-budget)	**35,120**	**320,778**	**316,602**	**320,819**
Surplus (+) or deficit (−)	**+56,430**	**+122,740**	**+69,242**	**+98,838**
(On-budget)	**+52,458**	**−951**	**−29,956**	**−24,759**
(Off-budget)	**+3,973**	**+123,691**	**+99,198**	**+123,597**

[1]These figures are based on the *Mid-Session Review of the FY 2000 Budget*, released by the Office of Management and Budget on June 28, 1999.

[2]Outlays have been increased and refunds of taxes decreased by $435 million for January 1999 through August 1999 to reflect additional reporting for payments where child care credits exceed the liability for tax.

[3]Outlays for the Department of Health and Human Services have been increased by $2 million and outlays for the Department of the Treasury — Other have subsequently been decreased by $2 million in September 1998 to reflect an adjustment by the Department of Health and Human Services.

[4]Outlays have been increased by $289 million in August 1998 to reflect an adjustment by the Small Business Administration.

.. No Transactions.

Note: Details may not add to totals due to rounding.

Figure 4-1

page under the heading, Social insurance and retirement receipts, can be found two listings; Total—FOASI trust fund (Social Security) of $383,559m and Total—FDI trust fund (Federal Disability Insurance) of $60,910m. The total of these two listing equals $444,469m.

In short, (Off-budget) items consist of FOASI (Social Security) and FDI (Federal disability Insurance). Figure 4-3 is a blow up of sections of Figures 4-1 and 4-2. Due to rounding, the last digit does not agree.

Social Security is by far the larger of the two numbers, representing about 86% of the total. For all practical purposes, it was the funds in Social Security that made it possible to claim a budget surplus. This claim was particularly misleading because Social Security is in extremely deep trouble now and it will worsen in the years ahead. As of the end of fiscal year 2001, Social Security had an **unfunded** liability of $14.442 trillion. It's unfunded because the government doesn't have sufficient receipts to pay the interest on this astronomical liability, assuming it was backed by debt securities as is the Public Debt. By 2038, the liability will be an inconceivable $143.319 trillion.

The Social Security taxes being collected in 1998 were in excess of the benefits being paid, which will be the case until 2025, according to the Social Security Administration. In 2026 benefit payments will exceed the taxes collected and in 2038 the cupboard will be completely bare.

In brief, the President's suggestion to move the surplus back to where it came from in the first place is nothing short of ludicrous. Did the President know where of he spoke or did he make the statement deliberately, knowing it would set well with the public?

Referring to Figure 4-2, without the Off-budget receipts the government would have shown a deficit of $346,558m in 1998 and $321,728m in 1999 . . . and this is only part of the story. Actually it was considerably worse as will be explained in the following paragraphs.

One would have thought that the President's error in suggesting the 1998 surplus be used to bolster Social Security would have drawn guffaws and ridicule from some of the not-so-friendlies in Congress.

Table 4. Receipts of the U.S. Government, September 1999 and Other Periods
($ millions)

Classification	This Month			Current Fiscal Year to Date			Prior Fiscal Year to Date		
	Gross Receipts	Refunds (Deduct)	Receipts	Gross Receipts	Refunds (Deduct)	Receipts	Gross Receipts	Refunds (Deduct)	Receipts
Individual income taxes:									
Withheld	[1]49,244			693,940			646,472		
Presidential Election Campaign Fund	1			61			63		
Other	[1]43,077			308,185			281,527		
Total—Individual income taxes	92,322	3,072	89,250	1,002,186	[1]122,706	879,480	928,063	99,476	828,587
Corporation income taxes	42,571	2,336	40,235	216,325	31,645	184,680	213,270	24,593	188,677
Social insurance and retirement receipts:									
Employment and general retirement:									
Federal old-age and survivors ins. trust fund:									
Federal Insurance Contributions Act taxes	[1]31,502	1,301	30,201	364,423	1,301	363,122	340,188	1,778	338,410
Self-Employment Contributions Act taxes	[1]3,528	......	3,528	20,437	......	20,437	20,379	......	20,379
Deposits by States	......	......	......	(* *)	......	(* *)	−5	......	−5
Total—FOASI trust fund	35,029	1,301	33,728	384,860	1,301	383,559	360,562	1,778	358,784
Federal disability insurance trust fund:									
Federal Insurance Contributions Act taxes	[1]5,008	206	4,803	57,891	206	57,685	54,076	293	53,783
Self-Employment Contributions Act taxes	[1]562	......	562	3,224	......	3,224	3,233	......	3,233
Deposits by States	......	......	......	(* *)	......	(* *)	−1	......	−1
Total—FDI trust fund	5,570	206	5,364	61,115	206	60,910	57,309	293	57,016
Federal hospital insurance trust fund:									
Federal Insurance Contributions Act taxes	[1]13,887	......	13,887	123,360	......	123,360	110,455	......	110,455
Self-Employment Contributions Act taxes	[1]1,475	......	1,475	8,520	......	8,520	9,029	......	9,029
Receipts from Railroad Retirement Board	......	......	......	388	......	388	381	......	381
Deposits by States	......	......	......	(* *)	......	(* *)	−2	......	−2
Total—FHI trust fund	15,362	......	15,362	132,268	......	132,268	119,863	......	119,863
Railroad retirement:									
Rail industry pension fund	180	1	179	2,633	4	2,629	2,599	16	2,583
Railroad Social Security equivalent benefit	160	1	160	1,517	3	1,514	1,782	12	1,770
Total—Employment and general retirement:	56,302	1,608	54,794	582,394	1,514	580,880	542,114	2,098	540,015
Unemployment insurance:									
Deposits by States	296	......	296	19,894	......	19,894	21,047	......	21,047
Federal Unemployment Tax Act taxes	39	3	36	6,650	175	6,475	6,479	111	6,369
Railroad unemployment taxes	(* *)	......	(* *)	111	......	111	68	......	68
Total—Unemployment insurance	335	3	332	26,655	175	26,480	27,595	111	27,484
Other retirement:									
Federal employees retirement — employee share	348	......	348	4,399	......	4,399	4,261	......	4,261
Non-federal employees retirement	8	......	8	73	......	73	74	......	74
Total—Other retirement	356	......	356	4,472	......	4,472	4,335	......	4,335
Total—Social insurance and retirement receipts	56,993	1,511	55,481	613,522	1,690	611,832	574,044	2,209	571,835
Excise taxes:									
Miscellaneous excise taxes	1,475	−234	1,710	20,760	524	20,236	23,008	714	22,294
Airport and airway trust fund	1,162	6	1,156	10,395	4	10,391	8,154	43	8,111
Highway trust fund	4,702	448	4,254	40,325	1,148	39,177	27,433	805	26,628
Black lung disability trust fund	48	......	48	596	......	596	636	......	636
Total—Excise taxes	7,387	220	7,167	72,076	1,676	70,399	59,231	1,562	57,669
Estate and gift taxes	2,348	54	2,294	28,386	603	27,782	24,631	555	24,076
Customs duties	1,788	61	1,727	19,486	1,150	18,336	19,689	1,392	18,297
Miscellaneous Receipts:									
Deposits of earnings by Federal Reserve Banks	2,789	......	2,789	25,917	......	25,917	24,540	......	24,540
Universal service fund	366	......	366	3,752	......	3,752	2,759	......	2,759
All other	1,096	9	1,087	5,848	740	5,108	5,073	46	5,027
Total — Miscellaneous receipts	4,251	9	4,242	35,516	740	34,777	32,371	46	32,325
Total — Receipts	207,659	7,262	200,396	1,987,496	160,211	1,827,285	1,851,299	129,834	1,721,465
Total — On-budget	167,059	5,756	161,304	1,541,521	158,704	1,382,817	1,433,429	127,763	1,305,666
Total — Off-budget	40,599	1,507	39,093	445,975	1,507	444,468	417,871	2,071	415,800

[1]Outlays have been increased and refunds of taxes decreased by $435 million for January 1999 through August 1999 to reflect additional reporting for payments where child care credits exceed the liability for tax.

[2]In accordance with the Social Security Act as amended; "Individual income taxes, Withheld" have been decreased and "Federal Insurance Contribution Act Taxes" correspondingly increased by $4,010 million to correct estimates for the quarter ending September 30, 1996. "Individual income taxes, Withheld" have also been increased and "Federal Insurance Contribution Act Taxes" correspondingly decreased by $42 million to correct estimates for calendar year 1996 and prior. "Individual income taxes, Other" have been decreased and "Self-Employment Contribution Act Taxes" correspondingly increased by $132 million to correct estimates for calendar year 1996 and prior.

... No Transactions.

(* *) Less than $500,000.

Note: Details may not add to totals due to rounding.

Figure 4-2

Table 3. Summary of Receipts and Outlays of the U.S. Government, September 1999 and Other Peric
[$ millions]

Classification	This Month	Current Fiscal Year to Date	Comparable Prior Period
Budget Receipts			
Individual income taxes	89,250	[2]879,480	828,587
Corporation income taxes	40,235	184,680	188,677
Social insurance and retirement receipts:			
Employment and general retirement (off-budget)	39,093	444,468	415,800
Employment and general retirement (on-budget)	15,701	136,411	124,215
Unemployment insurance	332	26,480	27,484
Other retirement	356	4,472	4,335
Excise taxes	7,167	70,399	57,669
Estate and gift taxes	2,294	27,782	24,076
Customs duties	1,727	18,336	18,297
Miscellaneous receipts	4,242	34,777	32,325
Total Receipts	**200,396**	**1,827,285**	**1,721,465**
(On-budget)	**161,304**	**1,382,817**	**1,305,666**
(Off-budget)	**39,093**	**444,468**	**415,800**
Budget Outlays			
Legislative Branch	201	2,621	2,600
Judicial Branch	317	3,793	3,463
Department of Agriculture	4,399	62,885	53,950
Department of Commerce	460	5,036	4,047
Department of Defense—Military	[illegible]	[illegible]	[illegible]

444,468

Table 4. Receipts of the U.S. Government, September 1999 and Other Periods
[$ millions]

Classification	This Month			Current Fiscal Year to Date			Pr
	Gross Receipts	Refunds (Deduct)	Receipts	Gross Receipts	Refunds (Deduct)	Receipts	Gro Rece
Individual income taxes:							
Withheld	[1]49,244			693,940			64
Presidential Election Campaign Fund	1			61			
Other	[1]43,077			308,185			281
Total—Individual income taxes	**92,322**	**3,072**	**89,250**	**1,002,186**	**[2]122,706**	**879,480**	**92**
Corporation income taxes	**42,571**	**2,336**	**40,235**	**216,325**	**31,645**	**184,680**	**21**
Social insurance and retirement receipts:							
Employment and general retirement:							
Federal old-age and survivors ins. trust fund:							
Federal Insurance Contributions Act taxes	[1]31,502	1,301	30,201	364,423	1,301	363,122	34
Self-Employment Contributions Act taxes	[1]3,528	......	3,528	20,437	......	20,437	2
Deposits by States	......	......	......	(* *)	......	(* *)	
Total—FOASI trust fund	35,029	1,301	33,728	384,860	1,301	383,559	36
Federal disability insurance trust fund:							
Federal Insurance Contributions Act taxes	[1]5,008	206	4,803	57,891	206	57,685	5
Self-Employment Contributions Act taxes	[1]562	......	562	3,224	......	3,224	
Deposits by States	......	......	......	(* *)	......	(* *)	
Total—FDI trust fund	5,570	206	5,364	61,115	206	60,910	5
Federal hospital insurance trust fund:							
Federal Insurance Contributions Act taxes	[1]13,887	......	13,887	123,360	......	123,360	11
Self-Employment Contributions Act taxes	[1]1,475	......	1,475	8,520	......	8,520	
Receipts from Railroad Retirement Board	......	......	......	388	......	388	
Deposits by States				(* *)		(* *)	

383,559 + 60,910 = 444,469

Figure 4-3

Such a response was notably absent; an indication that few if any of the members of Congress really understand budget details, but then again, how many people do you know who do. Some members of Congress stated they were in favor of using the so-called surplus to reduce taxes. This theme also plays well in Peoria and elsewhere.

In 1999 an article in one of the weekly issues of *Barron's*, a prestigious weekly financial publication, pointed out the source of the so-called surplus and the President's error. It drew little attention vis-à-vis Congressional voices loudly proclaiming where they wanted to spend it.

The Washington view on government accounting is unique to say the least. For example, funds collected from federal employees for their retirement, medical insurance, etc. are thrown into the pot with all other funds and referred to as receipts. Accounting wise, this is referred to as commingling, a practice unacceptable outside of Washington.

Washington established the Federal Employees Retirement Trust along with some 15 other trusts, one of which is Social Security, as separate entities whose funds were to be sacrosanct. Since the funds of these trusts are **only** usable for the designated purpose, they should not be involved in determining whether we have a surplus or a deficit. They are outside the government's day-to-day business. Table 4.1 reveals the surplus/deficit picture from 1960 through 2001, excluding the trust funds. For 1998, 1999, 2000 and 2001 the deficits were $725.5b, $738.0b, $667.6b and $816.6b respectively; combined some $2,947.6 trillion; certainly not a rosy picture. The numbers are frightening, yet the President and Congress voted a tax refund.

If the government's modus operandi were applied to the Ford Motor Company, retirement funds collected from Ford employees would be added to car sales and labeled income.

In Chapter 3, Table 3.2 reveals how the trust funds have grown over the years. In 1983, half of all of the receipts the government collected went into the trust funds. The tail began wagging the dog. By 1993, 53.8 cents out of every dollar collected was going into the trusts. However, the economic boom of the late 1990s bobbed the tail a bit and it dropped to 47.4 cents by 2001; likely to be only a temporary reprieve. On the out going side in 2001, 46.4 cents of every dollar the government spent was spent by the trusts. The Department of Defense, which is often accused of buying $200 hammers and toilet seats, had become a minor spender; they spent only 15.6 cents out of a dollar of outlays.

Along with Congress, the public for the most part finds details about our government's finances indecipherable and frustrating. With a few exceptions, the media apparently is no more knowledgeable than the general public since the President's ludicrous suggestion to use the surplus to bolster Social Security didn't make any headlines in the dailies.

Federal Surpluses & Deficits Excluding Trust Funds

The Government has set up more than 15 trusts for specific purposes, such as Social Security, Military Retirement, Unemployment, Federal Hospital Insurance, etc. As the name trust implies, funds in these trusts are for the purpose designated and no other—excluding these funds from the normal day-to-day government operations, such as defense, agriculture, etc. However, in reporting surpluses and deficits, government accounting does not separate the fiscal activities of the trusts from other operations—referred to as commingling in private business. It is the equivalent of the Ford Motor Company adding the money their employees contribute to retirement, health insurance, etc. to car sales and calling it total income, or using the government's term, receipts.

When trusts are excluded, the government reported surpluses were actually these deficits:

1998	$ 725.4 b
1999	738.0
2000	667.6
2001	816.6
	$ 2,947.6 b

Table 4.1 provides the fiscal surplus or deficits, excluding the trust funds, from 1960. The *Monthly Treasury Statement of Receipts and Outlays* lists trust fund receipts plus the interest earned, but not just receipts. Hence in Table 4.2 it was necessary to subtract one from the other to get only receipts. Also note in Table 4.1, the three month period in 1976 when the fiscal year was changed from July 1 through June 30 to October 1 through September 30 has been ignored.

Surpluses & Deficits Excluding Trust Funds

Year	All Receipts		Trusts Receipts				Surplus		
	$s m	Annual % Change	Income $s m	Interest $s m	Total $s m	Annual % Change	All Receipts less Trust Income $s m	Outlays $s m	Surplus + or Deficit - $s m
1960	92,492		19,005	1,337	20,342		73,487	92,191	-18,704
1961	94,388	2.0	22,168	1,414	23,582	15.9	72,220	97,723	-25,503
1962	99,676	5.6	22,857	1,433	24,290	3.0	76,819	108,821	-32,002
1963	106,560	6.9	26,212	1,477	27,689	14.0	80,348	111,316	-30,968
1964	112,613	5.7	26,887	1,613	28,500	2.9	85,726	118,528	-32,802
1965	116,817	3.7	27,460	1,770	29,230	2.6	89,357	118,228	-28,871
1966	130,835	12.0	31,089	1,908	32,997	12.9	99,746	134,532	-34,786
1967	149,600	14.3	40,660	2,275	42,935	30.1	108,940	157,464	-48,524
1968	153,700	2.7	42,042	2,674	44,716	4.1	111,658	178,134	-66,476
1969	187,800	22.2	48,910	3,099	52,009	16.3	138,890	183,640	-44,750
1970	193,743	3.2	55,464	3,936	59,400	14.2	138,279	195,649	-57,370
1971	188,392	-2.8	61,435	4,765	66,200	11.4	126,957	210,172	-83,215
1972	208,649	10.8	67,911	5,089	73,000	10.3	140,738	230,681	-89,943
1973	232,225	11.3	86,764	5,436	92,200	26.3	145,461	245,707	-100,246
1974	264,932	14.1	98,200	6,600	104,800	13.7	166,732	269,359	-102,627
1975	280,997	6.1	110,931	7,669	118,600	13.2	170,066	332,332	-162,266
1976	300,006	6.8	125,915	7,785	133,700	12.7	174,091	371,779	-197,688

Table 4.1a

Surpluses & Deficits Excluding Trust Funds

Year	All Receipts		Trusts Receipts				Surplus		
	$s m	Annual % Change	Income $s m	Interest $s m	Total $s m	Annual % Change	All Receipts less Trust Income $s m	Outlays $s m	Surplus + or Deficit - $s m
1977	357,762	19.3	144,669	8,131	152,800	14.3	213,093	409,203	-196,110
1978	399,561	11.7	159,349	8,651	168,000	9.9	240,212	458,729	-218,517
1979	463,302	16.0	179,649	9,951	189,600	12.9	283,653	503,464	-219,811
1980	517,112	11.6	201,854	12,046	213,900	12.8	315,258	590,920	-275,662
1981	599,272	15.9	225,603	13,797	239,400	11.9	373,669	678,209	-304,540
1982	617,766	3.1	252,409	15,991	268,400	12.1	365,357	745,706	-380,349
1983	600,563	-2.8	300,448	16,952	317,400	18.3	300,115	808,327	-508,212
1984	666,457	11.0	317,767	20,333	338,100	6.5	348,690	851,781	-503,091
1985	734,057	10.1	357,196	26,704	383,900	13.5	376,861	946,316	-569,455
1986	769,091	4.8	389,366	30,907	420,273	9.5	379,725	990,231	-610,506
1987	854,143	11.1	426,805	35,015	461,820	9.9	427,338	1,003,804	-576,466
1988	908,166	6.3	431,878	41,822	473,700	2.6	476,288	1,063,318	-587,030
1989	990,789	9.1	457,939	51,861	509,800	7.6	532,850	1,144,020	-611,170
1990	1,031,308	4.1	479,988	62,312	542,300	6.4	551,320	1,251,776	-700,456
1991	1,054,260	2.2	560,539	70,649	631,188	16.4	493,721	1,322,561	-828,840
1992	1,091,692	3.6	585,630	77,838	663,468	5.1	506,062	1,381,895	-875,833
1993	1,153,175	5.6	620,495	82,276	702,771	5.9	532,680	1,408,122	-875,442

Table 4.1b

Surpluses & Deficits Excluding Trust Funds

Year	All Receipts		Trusts Receipts				Surplus		
	$s m	Annual % Change	Income $s m	Interest $s m	Total $s m	Annual % Change	All Receipts less Trust Income $s m	Outlays $s m	Surplus + or Deficit - $s m
1994	1,257,187	9.0	640,481	85,698	726,179	3.3	616,706	1,460,557	-843,851
1995	1,350,576	7.4	670,105	93,176	763,281	5.1	680,471	1,514,389	-833,918
1996	1,452,763	7.6	718,204	98,029	816,233	6.9	734,559	1,560,094	-825,535
1997	1,578,977	8.7	754,476	104,992	859,468	5.3	824,501	1,601,595	-777,094
1998	1,721,421	9.0	795,503	113,830	909,333	5.8	925,918	1,651,383	-725,465
1999	1,827,285	6.1	860,733	118,634	979,367	7.7	966,552	1,704,545	-737,993
2000	2,025,038	10.8	904,620	128,911	1,033,531	5.5	1,120,418	1,788,045	-667,627
2001	1,990,203	-1.7	943,813	143,935	1,087,748	5.2	1,046,390	1,863,039	-816,649
2002	1,853,288	-6.9	977,839	153,195	1,131,034	4.0	875,449	2,011,808	-1,136,359
2003	1,782,317	-3.8	1,011,712	156,111	1,167,823	3.3	770,605	2,156,536	-1,385,931

Table 4.1c

THE PUBLIC DEBT

Always live within your income, even if you have to borrow money to do so.

— Josh Billings, Irish playwright and critic
Josh Billings' Encyclopedia of Wit & Wisdom

The principal focus of doomsayers in the past has erroneously been, and even today is, on the Public Debt. Table 5.1 provides a historical perspective of the debt beginning in 1960. The country's propensity to increasing its debt was illustrated in the 1980s. In the 6 years beginning in 1980 through 1985, a debt of $988.051 billion was accumulated by our government. It exceeded the total debt of $829.470 billion incurred in the 204 years prior to 1980, that is, since our country first came into being in 1776 through 1979. Further, note that projections by the Treasury Department for the years ahead are not down, but up. Even before the war on terrorism was initiated, the Treasury Department found the surpluses reducing the public debt dished out by the White House indigestible. As this book is being written, estimates for the cost of the war and its peripheral costs are very hazy, but high. Further, when the estimates are released they'll undoubtedly be on the low side.

With regard to the Public debt, there is a simple and valid yardstick which Fed-watchers monitor to alert them if the government becomes over extended. The measuring stick is how much of the government's useable receipts are needed to pay the interest on the Public Debt. Useable is the key word here. This is not a term or dollar number the government uses; they throw all dollars received into one pot and call it receipts. This commingles income taxes, money collected from federal employees for their retirement, military personnel retirement, railroad employees retirement, Social Security, medical and hospital insurance, interest earnings of trust funds, etc.

Table 5.1a

Public Debt

Year	Public Debt	Annual Increase	
(fiscal)	($s millions)	($s millions)	(percent)
1960	290,525		
1961	292,648	2,123	0.7
1962	302,928	10,280	3.5
1963	310,324	7,396	2.4
1964	316,059	5,735	1.8
1965	322,318	6,259	2.0
1966	328,498	6,180	1.9
1967	340,445	11,947	3.6
1968	368,685	28,240	8.3
1969	365,769	-2,916	-0.8
1970	380,921	15,152	4.1
1971	408,176	27,255	7.2
1972	435,936	27,760	6.8
1973	466,291	30,355	7.0
1974	483,893	17,602	3.8
1975	541,925	58,032	12.0
1976	628,970	87,045	16.1
1977	706,398	77,428	12.3
1978	776,602	70,204	9.9
1979	829,470	52,868	6.8
1980	909,050	79,580	9.6
1981	994,845	85,795	9.4
1982	1,137,345	142,500	14.3
1983	1,371,710	234,365	20.6
1984	1,564,657	192,947	14.1
1985	1,817,521	252,864	16.2
1986	2,120,629	303,108	16.7

Table 5.1b

Public Debt

Year	Public Debt	Annual Increase	
(fiscal)	($s millions)	($s millions)	(percent)
1987	2,346,125	225,496	10.6
1988	2,601,307	255,182	10.9
1989	2,868,039	266,732	10.3
1990	3,206,564	338,525	11.8
1991	3,598,498	391,934	12.2
1992	4,002,136	403,638	11.2
1993	4,351,416	349,280	8.7
1994	4,643,705	292,289	6.7
1995	4,921,018	277,313	6.0
1996	5,181,934	260,916	5.3
1997	5,369,707	187,773	3.6
1998	5,478,724	109,017	2.0
1999	5,614,934	136,210	2.5
2000	5,711,380	96,446	1.7
2001	5,807,463	96,083	1.7
2002	6,228,236	420,773	7.2
2003	6,783,231	554,995	8.9
2004	6,892,484	109,253	1.6 *
2005	7,206,865	314,381	4.6 *
2006	7,505,803	298,938	4.1 *
2007	7,803,313	297,510	4.0 *

* estimates from *The Budget forFiscal Year 2003*

It is the equivalent of the Ford Motor Company combining car sales with money collected from employees for medical and hospital insurance, retirement and calling it income. Such shenanigans are a "no-no" for private business; such funds must be segregated.

More than 15 government trust funds, refer to Chapter 3 for details, were set aside for specific purposes; their funds supposedly inviolate and not for the general business of government. The government trust funds acquire their money from special taxes, employee contributions and payroll deductions, interest, etc. and normally do not have a need to borrow money. As a specific example, why should a portion of the money a federal employee contributes for medical, hospital and life insurance, and retirement be used to pay the interest on the public debt.

In view of the preceding, trust fund receipts should be deducted from the government's total receipts to acquire a usable receipts number—useable receipts being defined as the funds available to cover normal government outlays. Based on useable receipts and interest expense it is easy to determine a vital indicator of the government's fiscal health—the percentage of receipts required to pay the interest expense on the public debt.

Useable receipts is derived by subtracting all trust fund receipts from total receipts—see Table 5.2. The right-most column of Table 5.2 shows the percentage of useable receipts required to pay interest expense each year since 1960. In the 1960s, the interest expense hovered around 12 percent of the useable receipts, a number you, I, or the government could easily live with. However, in the 1970s, interest as a percent of useable receipts began a slow but steady march upwards; the increase the first few years was innocuous enough not to raise any eyebrows. But by the close of the seventies, interest as percent of useable receipts was up nine percent. The first legitimate warning of fiscal trouble ahead came from William E. Simon, Secretary of the U.S. Treasury. In 1976 testimony before the House Appropriations Committee, he condemned the practice of describing legislated spending for future programs "future liabilities." In his words they were "present liabilities" payable in the future. Upon retirement as Secretary of the U.S. Treasury, he wrote a book summarizing his thinking, which in most respects is as valid today as

Table 5.2a

Interest Expenses versus Useable Receipts

Year (fiscal)	Receipts (total) $s m (1)	Trusts Receipts $s m (2)	Trusts Interest $s m (3)	Trusts Total (2) + (3) $s m (4)	Receipts (useable) $s m (5)	Interest Total $s m	Interest % of Receipts (useable)
1960	92,492	19,005	1,337	20,342	73,487	9,180	12.5
1961	94,388	22,168	1,414	23,582	72,220	8,957	12.4
1962	99,676	22,857	1,433	24,290	76,819	9,120	11.9
1963	106,560	26,212	1,477	27,689	80,348	9,895	12.3
1964	112,613	26,887	1,613	28,500	85,726	10,666	12.4
1965	116,817	27,460	1,770	29,230	89,357	11,430	12.8
1966	130,835	31,089	1,908	32,997	99,746	12,014	12.0
1967	149,600	40,660	2,275	42,935	108,940	13,400	12.3
1968	153,700	42,042	2,674	44,716	111,658	14,573	13.1
1969	187,800	48,910	3,099	52,009	138,890	16,588	11.9
1970	193,743	46,734	3,936	50,670	147,009	19,304	13.1
1971	188,392	61,428	4,765	66,193	126,964	20,959	16.5
1972	208,649	67,873	5,089	72,962	140,776	21,849	15.5
1973	232,225	86,773	5,436	92,209	145,452	24,167	16.6
1974	264,932	99,237	6,600	105,837	165,695	29,319	17.7
1975	280,997	111,893	7,669	119,562	169,104	32,665	19.3
1976	308,808	128,137	7,785	135,922	180,671	38,118	21.1
1977	367,782	145,353	8,131	153,484	222,429	41,900	18.8
1978	401,997	162,166	8,651	170,817	239,831	48,695	20.3
1979	465,955	181,818	9,951	191,769	284,137	59,837	21.1
1980	520,050	203,318	12,046	215,364	316,732	74,860	23.6
1981	599,272	230,829	13,797	244,626	368,443	95,589	25.9
1982	617,766	253,275	15,991	269,266	364,491	117,404	32.2
1983	600,562	300,718	16,952	317,670	299,844	128,813	43.0
1984	666,457	315,031	20,333	335,364	351,426	153,838	43.8
1985	734,057	368,421	26,047	394,468	365,636	178,945	48.9
1986	769,091	389,366	30,907	420,273	379,725	190,151	50.1
1987	854,143	426,805	35,015	461,820	427,338	195,390	45.7
1988	908,166	469,039	41,822	510,861	439,127	214,145	48.8

Table 5.2b

Interest Expenses versus Useable Receipts

Year (fiscal)	Receipts (total) $s m (1)	Trusts Receipts $s m (2)	Trusts Interest $s m (3)	Trusts Total (2) + (3) $s m (4)	Receipts (useable) $s m (5)	Interest Total $s m	Interest % of Receipts (useable)
1989	990,789	505,848	51,861	557,709	484,941	240,863	49.7
1990	1,031,308	528,262	62,312	590,574	503,046	264,853	52.6
1991	**1,054,260**	**560,539**	**70,649**	**631,188**	**493,721**	**286,022**	**57.9**
1992	1,091,692	585,630	77,838	663,468	506,062	292,330	57.8
1993	1,153,175	620,495	82,276	702,771	532,680	292,502	54.9
1994	1,257,187	640,481	85,698	726,179	616,706	296,278	48.0
1995	1,350,576	670,105	93,176	763,281	680,471	332,414	48.9
1996	1,452,763	718,204	98,029	816,233	734,559	343,955	46.8
1997	1,578,977	754,476	104,992	859,468	824,501	355,796	43.2
1998	1,721,421	795,495	113,838	909,333	925,926	363,824	39.3
1999	1,827,285	860,733	118,634	979,367	966,552	353,511	36.6
2000	2,025,038	904,620	128,911	1,033,531	1,120,418	362,118	32.3
2001	1,990,203	943,813	143,935	1,087,748	1,046,390	359,508	34.4
2002	1,853,288	977,839	153,195	1,131,034	875,449	332,537	38.0
2003	1,782,317	1,011,712	156,111	1,167,823	770,605	318,149	41.3

Note: Obviously the government trust funds should not be a source for paying interest expense on government debt securities which they hold as assets. Thus useable Receipts (column 5) equals total Receipts (column 1) less Trust receipts (column 2).

it was in 1978. Unfortunately his book, *A Time for Truth*, failed to stir up even a ripple of concern in Congress or the public.

Interest as a percent of useable receipts continued its steady rise during the 1980s. Fed watchers became more and more concerned. By 1990 they were very concerned. When the data for 1991 became available, they were terrified; 57.9 cents of every useable dollar the government collected was going to pay interest expense. It was a red alert. They believed government borrowing had pasted the point-of-no-return. That's when it becomes a Catch 22 situation—each year you must borrow more money just to pay last year's interest expense which increases your interest expense next year, and so on and on. Control of the debt is quickly lost. It becomes a runaway situation. The end is in sight.

Fortunately a good-time bubble came to the rescue and, by the close of fiscal year 2001, interest expense had declined to 34.4 cents of each useable dollar of receipts; still a scary number. As this is being written it appears that the good times have flown, slight-of-hand surpluses have evaporated, a recession is in progress, we're in a no-win terrorism war and a return to higher numbers is in the cards.

Although it is not at first obvious, who holds the public debt is a crucial factor in the event of a U.S. fiscal crisis. Refer to Tables 5.3, 5.4 and 5.5 for data on who holds the Public debt. A U.S. fiscal crisis would precipitate a loss of confidence in the dollar and inhibit the sale of our debt securities. The first to abandon ship are likely to be the foreign and international debt holders who own over 21% of the total public debt as of the close of fiscal 2000—see the appropriate column in Table 5.4. Finding buyers for even a small portion of the $1.2 trillion of debt securities held by fleeing foreigners would cause interest rates to skyrocket and likewise, the interest expense. The Public debt would become unmanageable.

Under normal circumstances, interest rates are typically controlled by the Federal Reserve Board. Contrary to popular belief, a free economy does not actually exist in the U.S. The Federal Reserve Board attempts and succeeds a great deal of the time, in controlling the economy by these actions.

1. Adjusting the Discount Rate—the interest rate charged member banks when they borrow funds short-term from the Federal Reserve to meet their reserve requirement.

Table 5.3

Ownership of Public Debt Securities

$s in billions

Year	Government Held		Privately Held		Total $s Public Debt
	$s	% of Total	$s	% of Total	
1990	1,026.0	31.7%	2,207.3	68.3%	3,233.3
1991	1,166.9	31.8%	2,498.4	68.2%	3,665.3
1992	1,282.4	31.6%	2,782.2	68.4%	4,064.6
1993	1,422.2	32.2%	2,989.3	67.8%	4,411.5
1994	1,562.8	33.3%	3,130.0	66.7%	4,692.8
1995	1,688.0	33.9%	3,286.0	66.1%	4,974.0
1996	1,831.6	35.1%	3,393.2	64.9%	5,224.8
1997	2,011.5	37.2%	3,401.6	62.8%	5,413.1
1998	2,213.0	40.0%	3,313.2	60.0%	5,526.2
1999	2,480.9	43.9%	3,175.4	56.1%	5,656.3
2000	2,737.9	48.3%	2,936.2	51.7%	5,674.2

Privately Held Public Debt Securiities

$s in billions

Fiscal Year	Depository Institutions	U.S. Savings Bonds	Pension Funds Private	Pension Funds State & Local Gov.	Insurance Companies	Mutual Funds	State & Local Govern-ments	Foreign and International $s	Foreign and International % of total Public Debt	Other	Total Privately Held $s	Total Privately Held % of total	Total Public Debt
1990	214.8	123.9	126.5	146.4	136.4	147.6	407.3	463.8	14.3%	440.7	2,207.3	68.3%	3,233.3
1991	251.7	135.4	126.2	140.2	171.4	199.5	430.2	506.3	13.8%	537.6	2,498.4	68.2%	3,665.3
1992	337.1	150.3	120.0	166.4	194.8	245.1	429.3	562.8	13.8%	576.5	2,782.2	68.4%	4,064.6
1993	366.2	169.1	125.1	188.7	229.4	283.9	434.0	619.1	14.0%	573.9	2,989.3	67.8%	4,411.5
1994	364.0	178.6	135.9	191.9	243.7	265.3	398.2	682.0	14.5%	670.4	3,130.0	66.7%	4,692.8
1995	330.8	183.5	141.4	193.0	245.2	272.6	304.3	820.4	16.5%	794.8	3,286.0	66.1%	4,974.0
1996	310.9	186.8	140.5	202.4	226.8	308.4	263.8	993.4	19.0%	760.2	3,393.2	64.9%	5,224.8
1997	292.8	186.2	141.6	219.7	187.3	311.4	237.7	1,230.5	22.7%	594.3	3,401.6	62.8%	5,413.1
1998	244.4	186.0	150.6	211.2	151.3	319.7	266.4	1,224.2	22.2%	559.4	3,313.2	60.0%	5,526.2
1999	239.9	186.2	164.2	217.3	130.6	338.3	271.6	1,281.4	22.7%	345.9	3,175.4	56.1%	5,656.3
2000	218.3	184.4	179.2	199.9	120.7	323.7	246.9	1,225.2	21.6%	237.9	2,936.2	51.7%	5,674.2

Table 5.4

Government Held Public Debt Securities

$s are in billions

Year (fiscal)	Government Held											Total Public Debt
	Trust Funds											
	Social Security	Federal Employees Retirement	Military Retirement	Railroad Retirement	Other Trust Funds	Total All Trust Funds	% of Total Public Debt	Govt. Depts., Agencies etc.	% of Total Public Debt	Total of all Govt.	% of Total Public Debt	
1996	499.4	401.8	116.9	17.1	339.2	1,374.4	26.3%	457.2	8.8%	1,831.6	35.1%	5,224.8
1997	567.4	430.8	126.0	19.2	358.7	1,502.2	27.8%	509.3	9.4%	2,011.5	37.2%	5,413.1
1998	653.3	460.6	133.8	21.8	386.1	1,655.7	30.0%	557.3	10.1%	2,213.0	40.0%	5,526.2
1999	762.3	491.9	141.3	24.3	449.5	1,869.2	33.0%	611.7	10.8%	2,480.9	43.9%	5,656.3
2000	893.5	523.2	149.3	24.8	518.3	2,109.2	37.2%	628.7	11.1%	2,737.9	48.3%	5,674.2

1) Government Held data is from appropriate Monthly Treasury Statements of Recipts and Outlays
2) Total Public Debt is from Table OFS-2, "Estimated Ownership of U.S. Treasury Securitiies," Office of Market Finance, Office of the Under Secretary for Domestic Finacne

Table 5.5

2. Purchasing and selling government securities in the open (secondary) market; their so called open market operations. By law they are prohibited from purchasing Government securities directly from the Government.
3. Controlling the minimum reserve requirements banks must maintain.

The Federal Reserve's most publicly visible action is changing the Discount Rate; it is lowered to energize the economy or raised to keep inflation in check.

Another factor of the Public Debt is it's average maturity; five years, nine months as of March 31, 2001 and trending down. In December 1975 it hit a low of two years, five months. These numbers are probably lower than most people would guess. The short maturity of the debt is a factor in establishing the current interest rate and likewise the annual interest paid by the government. It also explains how the government's fiscal health can quickly improve or degrade in a matter of just a few years. Further it reveals why the Federal Reserve exerts constant pressure to keep interest rates low, a bonus to the government and business, but a severe detriment to savers.

ACCOUNTING ABERRATIONS

No one surpasses the government when it comes to numbers.

— ***R. Earl Hadady***

Various government data sources used in this book will be found in the References section. Practically all of the data was obtained from the four government organizations which manage the specific trusts funds covered herein and appropriate issues of the Treasury Department's, *Monthly Treasury Statement of Receipts and Outlays*. Members of these organizations were particularly helpful and cooperative.

It should be noted the same data from different sources will be at variance. Examples of these differences will be found in Table 6.1 which compares data from the Treasury Department's publications, *Monthly Treasury Statement of Receipts and Outlays*, versus data from the four organizations that manage the trust funds. Percentage wise the differences were small except in the case of the Railroad Retirement Trust, the smallest of the four trusts. This trust organization does not provide a total of the four different accounts which make up this trust—original Railroad Retirement Account, Supplemental Account, Dual Benefits Account, and the Social Security Equivalent Benefits Account. Numbers used herein closely matched data supplied by the Railroad Retirement trust, but both were significantly lower than the Treasury Department's numbers.

For analytical purposes herein, the data from trust funds was used in most instances; it was presumed to be more accurate and presented a less pessimistic outlook of a catastrophic situation. Further, to be as optimistic as possible, cost-of-living-adjustments (COLAs) were ignored.

It is interesting that the dollar numbers from the Treasury Department exceed those from three of the trust funds, an indication it's not likely to be a coincidence. Possibly special Treasury Department expenses for the trusts are included in their outlay numbers. On the other hand the Military Retirement trust data corresponded exactly with Treasury data. As stated above, it's worthy of note, but of no consequence in achieving the objective of this book—determining approximate current and future benefit liabilities of Social Security and the three retirement trust funds.

Table 6.1

Data Comparisons

Treasury Department versus Trust Fund Group

The Treasury Department's data in the following tables has been acquired from copies of the *Monthly Treasury Statement of Receipts and Outlays*, Table 8. The Trust Fund Group data was acquired directly from group managing the trust fund—their data was assumed to be more accurate and was used herein to calculate the liabilities. In general, this approach reduces the liabilities.

Securities Held as Investments

(At close of fiscal year 2000 - $s in millions)

	Social Security	Federal Employees Retirement	Military Retirement	Railroad Retirement
Treasury Dept.	$ 901,597*	$ 523,200	$ 149,348	$ 24,823
Trust Fund Group	893,003	512,046	149,348	19,551
Difference from Group	$ 8, 594	$ 11,154	$ 0	$ 5,272
% Difference from Group	0.9%	2.2%	0%	27.0%

*As of December 31, 2000— for comparison with Social Security which operates on a calendar year basis.

Outlays for the Month

(Last month of fiscal year 2000 - $s in millions)

	Social Security	Federal Employees Retirement	Military Retirement	Railroad Retirement
Treasury Department	$29,864†	$3,834	$2,740	$ 686
Trust Fund Group	29,392†	3,767	2,734	1,003
Difference from Group	$472	$ 67	$ 6	$-217
% Difference from Group	1.6%	1.8%	0.2%	-21.6%
Treasury Department†	$29,452	$3,815	$2,734	$ 703

† For comparative purposes this is the average of the 12 months of 2000.

SOCIAL SECURITY FALLACIES

First payments is what made us think we were prosperous, and the other nineteen is what showed us we were broke.
— *Will Rogers*

Depending upon who is doing the defining, Social Security is many different things. For example:

It (Social Security) runs a complicated insurance program, biggest in the world, . . .[1]
— *Robert M Ball*
Commissioner of Social Security 1962-1973

It (Social Security) serves as the primary source of retirement income for most Americans.[1]
— *Eugene I. Lehrmann, President*
American Association of Retired Persons 1995

These programs (Social Security) are insurance, not welfare.[1]
— *Alan Brinkley, Professor*
Columbia University

Social Security has never been an entitlement program; it is an 'earned-right' program, and it's Congress's job to do whatever is necessary to deliver it—today, tomorrow, and in the 21st century.[1]
— *Bill Archer (R-Texas)*
U.S. Representative

Contrary to popular myth, Social Security is not a savings program. It is a strong and noble commitment by the generations who are currently working to pay for the retirement of those who are not, with the expectation that the generations to come will do the same when they retire.[1]
— *Bob Kerrey (D-Nebraska)*
U.S. Senator

Clearly Bill Archer, former Representative from Texas, and Bob Kerrey, former Senator from Nebraska, are at odds on what Social Security is. Archer believes it is a savings program in which the participants are entitled to get their money back. Kerrey's definition of Social Security fits the dictionary's definition of a Ponzi scheme. A Ponzi is defined in the dictionary as an investment swindle in which early investors are paid off with money put up by later investors. The name is derived from Charles A. Ponzi, an Italian-born American swindler. And there are those who think it is an insurance program, for example, none other than the former head of Social Security. Maybe they forgot to send out our policies. Since Washington can't agree on what Social Security is, it's no wonder the populace is confused.

There is even some argument over the meaning of "entitlement." Just for the record, Merriam Webster defines entitlement as, "a right to benefits specified by law or contract; also, a government program providing benefits to members of a specified group." Note that this definition does not exclude an "annuity" contract as an "entitlement."

However, arguing over what Social Security is and what the intent of the program was when it was established in 1935 is pointless.

Although not intended as such when it was created during the Great Depression in 1935, today in the real world Social Security has become a retirement annuity for about two out of three older Americans.

> *Social Security is the major source of income for two thirds of older Americans and virtually the only source of income for one-third of older Americans. In a real sense, Social Security is the most important income security program in American history.*[3]
>
> — ***Acting Commissioner Halter***

In view of the growth of the Social Security operations, Congress voted to separate it from the Department of Health and Human Services in 1994. On March 31, 1995, Social Security officially became an independent government agency. As of the end of calendar year 2000, there were approximately 38 million persons receiving Social Security benefits.

There are some people, including Representative Bill Archer, that believe the typical Social Security recipient receives much more than he/she ever puts into the fund. Because the Social Security rolls include some 30% disabled persons, widows, and surviving children, it is impossible to categorically answer this question. Only a study by the Social Security Administration can provide an accurate answer. However, the information and data included in an April 15, 1996 article "No Bonanza," written by the author for *Barron's*, covers a hypothetical John Doe, his payments into Social Security and his benefits upon retirement. A summary of the article follows; details will be found in Appendix D.

In the example, John Doe, a white American male, was born in 1930. In January 1950, at age 20, Doe began work at a starting salary of $500 a month. He received annual salary increases of 4% during his career and worked until retirement in January 1995, when he reached the age of 65. Based on the Social Security tax-rate schedule, the total payments by Doe and his employer amounted to $64,224 during the 45-year period. Had Doe been able to put his payments into 52-week T-bills, his payments would have grown to $159,956 which would have provided him an income of around $1440 each month for the remainder of his life.

For comparative purposes, John Doe's history-of-earnings were furnished to Social Security. Figuring Social Security payments are quite complicated because they adjust annual earnings for inflation and average a large number of the annual earnings, etc. However, Social Security generously took the time and effort to determine the amount of Doe's monthly benefits and other pertinent information[2]. A comment aside, the author wishes to point out he found the Social Security personnel very courteous and cooperative. Their calculations showed John Doe would receive benefits of about $1078 per month.

Social Security was no bargain for John Doe, nor will it be for anyone having a similar work history. The benefits fall about 30% or more short of what an enrollee might receive if allowed to contribute to a retirement pool invested in government securities. If you believe a Social Security participant is entitled to earn interest on his or her contributions during the many years prior to receiving benefits, then it's really beside the point as to whether Social Security was or was not intended as a retirement fund. A participant with a work history similar to John Doe's is entitled to all the Social Security benefits he or she gets and then some.

The above return from Social Security may be at odds with what you have heard elsewhere or read some other place. For instance:

> *"Many of today's retirees don't realize how fast they get their money back plus interest—once they start collecting benefits. An average earner who retires this year at 65 will get back all his or her contributions plus interest in eight years. For workers who always earned the maximum the payback time is abut 11 years—still a very good return given that at 65 the average man is now expected to live past age 80, the average woman to almost 85."*[1]
>
> — ***Bill Archer (R-Texas)***
> *U.S. Representative*

There were a couple of things that Congressman Archer failed to mention. First and most importantly, he apparently only included half of the tax that was paid into Social Security for the retiree—only what an employed retiree paid. The tax contributed by the employer for the employee was left out. Possibly he thinks the money contributed by the employer belongs to the government. How about the self-employed person who pays both the employee's and the employer's tax—no mention of this. Perhaps he thinks that the Social Security program should differ from pension plans in private industry in which the employer's contribution and earnings become part of the employee's retirement package. The company's contribution is always considered as belonging to the employee once the employee has worked a prerequisite number of years. After the prescribed period, the employee is said to have a fully vested interest in what the employer had contributed. When you double the number of years that Congressman Archer specified, from 8 to 16 and 11 to 22, where is this bargain he talks about? The recipient may have become deceased before then. A second point, Congressman Archer did not specify the interest rate, how it was applied, and over what period.

In short, it is wise to question any and all similar statements from government officials unless it is backed with specifics and data.

Whether or not the John Doe example represents millions of the participants in the program can only be answered by an extensive Social Security Administration study of the records. However, John Doe is probably a good example of how many persons view their relationship with Social Security.

In the John Doe example, his 1979 annual earnings fell below the maximum income on which Social Security collects taxes. The point, for someone who's annual earning's always exceeded the maximum taxed by Social Security, the payments would have been more than Doe's; the total taxes would have amounted to $96,486.22 versus $64,223.94 to be exact. Invested in 52 week T-Bills, these payments would have grown to $206,253.05 during the 45 year period. This nest egg would provide an income for life of about $1850 per month. As of 1996, Social Security's payments for an individual topped out at $1199 per month for a retiree at 65. Again, in this instance, Social Security benefits fall considerably short of what an enrollee is entitled to receive.

For retirees at 66, 67, 68, 69 and 70, Social Security benefits as of 1996 maxed out at about $1259, $1322, $1388, $1457 and $1530 respectively. But remember, the retiree's life expectancy has shortened. Also recall that an annuity grows the fastest during the last years because the principal is larger. Hence, an annuity would probably be paying at least $1000 more per month than Social Security for a person retiring at age 70.

Any way you slice it, Social Security is no bonanza.

References:

1) AARP Focus on Social Security, July-August 1995, Modern Maturity

2) March 4, 1995 Fax – sent by Tom Margenao at the Social Security Administration, Press Office in Baltimore, MD to R. Earl Hadady

3) Social Security News Release, Monday March 19, 2001

THE POINT OF NO RETURN

We can't cross a bridge until we come to it; but I always like to lay down a pontoon ahead of time
— Bernard Mannes Baruch (1870–1965)
US financier & government advisor

In the present context, the point of no return means there is no going back; no returning to a prior position and starting over; proceeding ahead is the only option.

A person, corporation, or the government reaches the fiscal point of no return when meeting financial commitments is no longer possible. The only option is to restructure the commitments.

As of the close of fiscal 1981, the government was bordering on the point of no return. The Public Debt was a mere $0.99 trillion; total Social Security present and future benefit liabilities were $2.74 trillion; the three retirement trust fund liabilities were small. Short term interest rates were quite high in 1981 due to excessive defense spending, but if Social Security had been fully funded by issuing 4% interest bearing government securities, the government could have probably squeaked by. Useable receipts for the year were $368 billion while on-going interest expense was $96 billion. Fully funding Social Security's $2,835 billion liabilities with 4% interest bearing short-term securities would have resulted in an additional interest expense of $113 billion. This would have brought the total interest expense to $209 billion, some 56% of the $368 billion in useable receipts for that year. An interest rate higher than 4% would have been desirable, but by using short-term securities, changes could have been made in the following decade to correct any deficiencies. An interest expense of 56% is a very high percentage of useable receipts and in itself puts government finances on the edge of a Catch 22 situation, that is, each year additional borrowing is required to pay last year's interest expense—when this occurs, insolvency is just around the corner.

Some ten years later in 1991, interest expense as a percent of useable receipts reached the scary level of 57.9% and many doomsayers were of the opinion that we had had it. Fortunately in the following years, good times began to roll and increased government receipts improved the situation.

Looking at the close of fiscal 2001, useable receipts were $1,046 billion. Ongoing interest expense was $360 billion. Total current and future benefit liabilities of Social Security were $14,442 billion. Again assuming fully funding Social Security with 4% interest bearing government securities, the additional interest expense would be $578 billion, bringing the total interest expense to $938 billion. An interest expense of $938 billion would have represented 89.7% of the $1,046 billion in useable receipts, clearly an impossible option. Further, this didn't include fully funding the growing liabilities of the Federal Employees Retirement Trust, the Military Retirement Trust and the Railroad Retirement Trust of $969 billion, $934 billion and $526 billion respectively.

As a temporary means of getting by, the government refers to these four trusts as unfunded liabilities. Trust fund receipts are held a relatively short time and hence produce little earnings before they must be used to pay benefits. Paying benefits with receipts, which are not held long enough to generate sufficient earnings to pay the benefits, is generally referred to as a Ponzi, a procedure ruled illegal for private business.

Former Secretary of the Treasury, William Simon, correctly identified the government's underlying problem in 1976. Congress simply wasn't bright enough to understand what he was saying and do something about it. In his 1978 book, *A Time for Truth*, he stated, "In 1976, in testimony before the House Appropriations Committee, I condemned the practice of describing legislated spending for future programs 'future liabilities.' That, I said was deceptive formulation, they are present liabilities payable in the future, and the budget should make this clear by using the accrual accounting method."

And where to from here. As can be seen in the Appendix C, Social Security liabilities are growing faster than the government's receipts. Hence, the quicker the liabilities are restructured, the better; the damage will be minimized. Although this sounds like a simple solution, keep in mind the words of the Social Security Acting Commissioner William Halter, "Social Security is the major source of income for two-thirds of older Americans and virtually the only source of income for one-third of older Americans." Any restructuring of benefits to two-thirds of older Americans is, at best, going to have a permanent and debilitating effect on American society.

And don't be taken in by the political subterfuge that Congress is considering allowing individuals to invest part of taxes they would normally pay into Social Security into a private retirement fund. This would only exacerbate Social Security's problems. Because it is a Ponzi operation, in which funds do not remain in the trust long enough to earn sufficient interest to pay the benefits, it typically takes about three persons paying into Social Security to pay benefits to one person. Consequently, reducing the number of people paying or partially paying into Social Security would only make matters worse.

And what about world confidence in the U.S. dollar. The only thing you can count on is it wouldn't be good. Over 20 percent of the U.S. Public Debt is held by foreign and international money lenders—see Chapter 5. If these lenders withdrew a significant amount of their funds, the Treasury Department would be forced to raise interest rates to attract new investors. Higher interest rates are bad news for the government because a larger portion of receipts would have to go into paying interest expense, and it is already high, 34.4% in 2001.

HISTORIC RISE AND FALL OF GREAT NATIONS

The only thing we learn from history, it has been said, is that men never learn from history

— author unknown

Prior to summing up the U.S. fiscal situation based on data presented in earlier chapters and the appendices, a brief look at the historic rise and fall of great nations of the world can provide an overall perspective.

History reveals, in most instances, both the rise and fall of great nations of the past are so gradual that they are indiscernible until they becomes de facto. It is rather obvious that if the beginning of a country's decline from its zenith were clearly and generally recognized, its citizens would take appropriate action to reverse the trend. Possibly a country's rise to power develops a mind set in its citizens of arrogance, infallibility, and superiority that precludes recognizing the beginning of the end.

Considerable effort has gone into researching the rise and fall of great nations. One such effort is *The Fate of Empires and Search for Survival* by Sir John Bagot Glubb, a 46 page booklet published by William Blackwood & Sons Ltd. of Edinburgh, Scotland in 1978. Sir Glubb concluded that great civilizations tend to have a life cycle of about 250 years. His life estimates of a number of nations are shown in Table 9.1. It is interesting to note that the life spans of nations are apparently related to human behavior and **not** a function of the technological environment.

Sir Glubb concluded the following about great nations:

1. Average length of world greatness is roughly 250 years.

Table 9.1

Duration of Great Nations

Nation	Note	Rose	Fell	Duration (years)
Assyria		859 B.C.	612 B.C.	247
Persia	(1)	538 "	330 "	208
Greece	(2)	331 "	100 "	231
Roman Republic		260 "	27 "	233
Roman Empire		27 "	180 A.D.	207
Arab Empire		634 A.D.	880 "	246
Mameluke Empire		1250 "	1517 "	267
Ottoman Empire		1320 "	1570 "	250
Spain		1500 "	1750 "	250
Romanov Russia		1682 "	1916 "	234
Britain		1700 "	1950 "	250

Notes: (1) Cyrus and his descendants
(2) Alexander and his successors

Table and data courtesey of Sir John Glubb and his *The Fate of Empires and Search for Survival*, published by William Blackwood & Sons Ltd, Edinburgh Scotland, 1978

2. The rise and fall can be divided into a sequence of periods:
   Pioneering
   Conquests
   Commerce
   Affluence
   Intellect
   Decadence
3. The period of Decadence is marked by:
   Defensiveness
   Pessimism
   Materialism
   Frivolity
   Influx of Foreigners
   The Welfare State
   Weakening of Religion
4. Decadence is due to:
   Wealth
   Power
   Selfishness
   Love of money
   Loss of sense of duty
5. The life histories are amazingly similar, and are primarily the result of internal human factors
6. Their falls are diverse because they are largely the result of external causes.

Sir Glubb's numbers provide a reasonable fit with the numbers developed in this work. Applying the 250 year life cycle to the U.S. would place the end of the cycle at approximately 2026. Allowing a plus or minus 50-year variation from the 250 norm, the life cycle of the U.S. could have ended as early as 1976 or be as late as 2076. 1976 was the year the U.S. began to commit fiscal suicide and approximately the last year the unfunded trusts could have been funded. Later years agree with the period during which the U.S. will have to come to grips with restructuring Social Security, the Federal Employees Retirement Trust, the Military Retirement Trust and the Railroad Retirement Trust.

Further, Sir Glubb cited some historical conditions similar to the present.

> The Roman mob, we have seen, demanded free meals and public games. Gladiatorial shows, chariot races and athletic events were their passion. In the Byzantine Empire the rivalries of the Greens and Blues in the Hippodrome attained the importance of a major crisis.
>
> The historians commented bitterly on the extraordinary influence acquired by popular singers over young people, resulting in a decline in sexual morality.
>
> Several khalifs issued orders banning 'pop' singers from the capital, but within a few years they always returned.

The Rise and Decline of Nations by Professor Mancur Olson, a 273 page book published by Yale University Press in 1982 is another relevant work. Professor Olson developed a theory built on arguments advanced in a prior work of his, *The Logic of Collective Action.* His thesis is, "the behavior of individuals and firms in stable societies leads to the formation of dense networks of collusive, cartelistic, and lobbying organizations that make economies less efficient and dynamic and polities less governable." The professor went on to state, "The longer a society goes without an upheaval, the more powerful such organizations become—and the more they slow down economic expansion. Societies in which these narrow interest groups have been destroyed—by war or revolution, for example—enjoy the greatest gains in growth."

It would seem that Professor Olson's conclusions reflect Thomas Jefferson's thinking when he said, "A little rebellion now and then … is a medicine necessary for the sound health of government."

The basis of Professor Olson thesis is an apt description of the Washington crowd nowadays which, in addition to the White House and Congress, includes PACs (Political Action Committee), private and corporate lobbyists, influence peddlers etc.

A third work dealing with the rise and fall of nations, *The Rise and Fall of the Great Powers*, was written by Paul Kennedy, Dilworth

Professor of History at Yale University. This work of 677 pages was published by Random House in 1987. In his book he stated:

> Although the United states is at present still in a class of its own economically and perhaps even militarily, it cannot avoid confronting the two great tests which challenge the longevity of every major power that occupies the 'number one' position in world affairs: whether, in the military/strategical realm, it can preserve a reasonable balance between the nation's perceived defense requirements and the means it possesses to maintain those commitments; and whether, as an intimately related point, it can preserve the technological and economic bases of its power from relative erosion in the face of the ever-shifting patterns of global production. This test of American abilities will be the greater because it, like imperial Spain around 1600 or the British Empire around 1900, is the inheritor of a vast array of strategical commitments which had been made decades earlier, when the nation's political, economic, and military capacity to influence world affairs seemed so much more assured. In consequence, the United States now runs the risk, so familiar to historians of the rise and fall of previous Great Powers, of what might roughly be called 'imperial overstretch': that is to say, decision makers in Washington must face the awkward and enduring fact that the sum total of the United State's global interest and obligations is nowadays far larger than the country's power to defend them all simultaneously.

Professor Kennedy's book made the New York Times best-selling non-fiction list for 17 weeks. His subject also provoked a surprisingly strong reaction, specifically among Washington officials. A May 31, 1988 Los Angeles Times article reported, "Officials Upset by Book Saying U.S. Is Courting Decline." Washington officials have never been known to take criticism too kindly; their mind set "it can't happen to us" didn't make it any easier.

Although these three books didn't deal specifically with the Public Debt of the U.S., they were published shortly after 1970 when the debt, as a percent of receipts, began to steadily rise from an acceptable level where it had been residing for well over ten years.

Notable Criticisms of U.S. Fiscal Conduct

Legitimate criticisms of the conduct, policies and actions of the U.S. don't set well with its typical citizen, not to mention the Washington crowd. To even suggest that the U.S. may be faltering as the world leader creates groundless indignation. Citizens are reluctant to consider: 1) we may have too many irons in the fire at home and abroad, 2) we may have bitten off more than we can chew and 3) the capabilities of the U.S. are indeed limited and 4) we can over commit fiscally and physically. One such major lesson, our war in Vietnam, seems to have been forgotten.

Typically criticisms are simply ignored until they are forgotten by Washington and the public. The following are some examples:

> *In 1976, in testimony before the House Appropriations Committee, I condemned the practice of describing legislated spending for future programs 'future liabilities.' that, I said, was deceptive formulation; they are present liabilities, payable in the future, and the budget should make this clear by using the accrual accounting method*
>
> ***— A Time for Truth, by William E. Simon***
> *(former Secretary of the Treasury 1947-1977).*
> *McGraw-Hill Book Company, 1978, p. 100.*

> *The United States has been living in a 'false paradise' in which we have grown dependent on heavy flows of foreign capital to keep our standard of living rising.*
>
> *— **Paul A. Volcker,** former Chairman Federal Reserve System. Testimony before the House Ways and Means Subcommittee, September 24, 1986*

> *I agree with Hadady that the interest bill on the federal debt ought to put Uncle Sam in the intensive care unit. I'm getting little nervous that so many people still can't see how sick the old gentleman is.*
>
> *— **Lee Iacocca,** Former Chairman of the Board, Chief Executive Officer Chrysler Corporation.*

A 1986 endorsement of How Sick is Uncle Sam
by R. Earl Hadady, Key Books Press, 1986

Our government is in serious financial trouble. Hadady and I do not agree completely on the causes of explosive government spending. However, the author, in my view correctly points out that our nation's overhead must be managed as a well-run business would manage its own expenses—declining as a percentage of the overall operation as that operation expands.

— ***J. Peter Grace,*** *Chairman of the President's Private Sector survey on Cost Control (known as the Grace Commission) A1986 endorsement of How Sick is Uncle Sam, by R. Earl Hadady, Key Books Press, 1986*

One thing is certain. At some point global investors will lose confidence in our easy dollars and debt-financed prosperity, and then the chicken will come home to roost.

— ***The Triumph of Politics, Why the Reagan Revolution Failed, by David A. Stockman,*** *Harper & Row, Publishers, 1986, p.379*

The deficit is a malignant force in our economy. Allowing it to fester would court a dangerous erosion of our economic strength and a potentially significant deterioration in our real standard of living.

— ***Alan Greenspan,***
Chairman, Federal Reserve System
Statement before the Committee on Finance,
U.S. Senate, March 24, 1993, p. 473

Social Security is the major source of income for two-thirds of older Americans and virtually the only source of income for one-third of older Americans. We should utilize this period of Social Security surpluses to fashion the legislative changes necessary to extend the solvency of the system for future generations.

— ***William A. Halter,***
Social Security Acting Commissioner.
Social Security News Release,
March 19, 2001

SUMMING UP

The obvious is that which is never seen until someone expresses it simply.

— Kahlil Gibran

The deteriorating fiscal condition of the U.S. Government can be easily understood by examining the trust funds—how these funds are commingled with other receipts—how these funds generate fictitious surpluses. The government's problem will become clear by analyzing the liabilities of four key trusts and why they are unfunded—Social Security, Federal Employees Retirement, Military Retirement, and Railroad Retirement. In this work, the hard numbers developed from data provided by federal sources sets it apart from previous work dealing with the government's fiscal problems.

The Trust Funds

Four trust funds are the fiscal Achilles heel of the U.S. government. Figure 3-1 in Chapter 3 is a copy of Table 8 in the *U.S. Final Monthly Treasury Statement of Receipts and Outlays*, dated September 30, 2001. This statement covers the month of September and summarizes fiscal year 2001. This table lists some 15 different specific trusts and one listing which incorporates several minor trusts. These trusts were established during past years by Congress to provide funds for a specific purpose and only that purpose, such as Social Security, Federal Employees Retirement, Hospital Insurance, Veterans Life Insurance, Unemployment, etc. In most instances, money for these funds is collected from workers in industry, federal employees, railroad workers, etc. The funds in these trusts are for an explicit use and not for day-to-day government operations. Although these trust funds occupy only a half-page in the 31 page monthly

Treasury Statement, they have become the tail that wags the dog. Many will be surprised to learn that each year during the period 1985 through 1994, more than 50 cents of every dollar collected by the government was slated for the trust funds; in 2001 it was 47.4 cents. On the other hand, trust fund outlays were rising each year and by 2001 they accounted for 46.4 cents of every dollar the government spent. The remainder of the government, the White House, Congress, the Judicial Branch, all Departments (including the Department of Defense), and about 60 independent establishments and government corporations had to divvy up the remaining 54 cents. The Department of Defense only got 15.6 cents of each dollar.

See Tables 3.2 and 3.3 in Chapter 3 for a tabulation of annual trust receipts and outlays compared to all receipts and outlays beginning in 1982. All data are from the *Monthly Treasury Department Statements of Receipts and Outlays.*

Of passing interest, in 1992 the GAO (Government Accounting Office), which is the Congressional watchdog, pointed out that the trust fund receipts listed in the monthly Treasury statements were only estimates. It seems unbelievable that in the case of Social Security, this sloppiness had been going on since 1937. Subsequently the Internal Revenue Service corrected this minor oversight. In short, the government accounting system doesn't meet the standards they require of private industry, or any industry for that matter. Other accounting liberties by Big Sam will be described later in this chapter. His Enron accounting practices needs a complete overhaul; an overhaul supervised by several outside respected major accounting firms, Arthur Andersen excepted.

Commingling Funds

Although the Internal Revenue Service stopped **estimating** receipts and started providing **actual** receipts for each trust fund in 1992, government accounting is certainly no role model, far from it. The government continues to commingle trust fund receipts with receipts from all other sources, such as personal and corporate income taxes, etc. to determine government surpluses and deficits. This is an absolute "no-no" in private business. It would be the equivalent of the

Ford Motor Company taking money collected from employees for retirement and medical insurance, adding it to car sales and calling it income. This very deceptive government practice is covered in the next section.

The legality of this government accounting practice is also questionable, since trust funds are only usable for the purpose specified, not the general purpose of government. It would be interesting to get a Supreme Court ruling on this misleading procedure, but the odds on this happening lie somewhere between slim and none.

Smoke, Mirrors and Surpluses

Money in most of the trust funds has been collected from workers in private industry and the government for retirement, life insurance, hospital insurance, medical insurance and similar purposes. Hence, these funds are not involved in the normal day-to-day operations of the government. Thus to correctly determine the government's annual surplus or deficit, the trust funds should be excluded, just as they would be in private industry. The government wouldn't allow the Ford Motor Company to commingle money from car sales with money collected from employees for retirement, medical insurance, etc. and call it income; referred to by some as Enron accounting.

Table 4.1 in Chapter 4 provides data since 1960 for the government's total receipts, trust receipts, useable receipts (total receipts less trust receipts), outlays and the surplus/deficit (useable receipts less outlays). The last surplus was sometime prior to 1960. As can be seen the deficits have actually been rising in recent years. The surpluses hailed by Presidents Clinton and Bush and the Congress in 1998, 1999 and 2000 were actually deficits of $725, $738 and $668 billion, respectively; in each case approaching a trillion dollars—and in 2001 it was nearly one trillion, $817 billion.

Consider these facts: President Clinton wanted to put the fictitious 1998 surplus into Social Security, which is where it came from in the first place; Congress wanted to spend it to garner votes for guess who; President Bush, who wasn't familiar with Social Security, wanted to use the surpluses to reduce taxes. Musing on these facts, the very,

very best face that can be put on the White House and Congress boils down to fiscal incompetence. And it might even be worse. When the results of fiscal 1998 became known in November 1998, *Barron's*, the prestigious weekly business publication, was quick to point out the fallacy of the ballyhooed 1998 surplus—meanwhile the White House and Congress were too busy pitching political hay to pay attention to the real world.

Retirement Trust Funds are Unfunded

The crux of the problem is four trust funds: Social Security (the biggie), Federal Employees Retirement, Military Retirement, and Railroad Retirement. With the exception of the Federal Employees Retirement Trust, benefits are currently being paid with recent receipts, a modus operandi commonly referred to as a Ponzi; an operation which is inherently faulty and will eventually fail. Avid Ponzi followers are limited to those who believe in chain letters and the tooth fairy. The fiscal modus-operandi of these trusts is inherently defective because receipts don't hang around anywhere near long enough to generate sufficient earnings to pay the benefits. Further, this operating scheme is illegal outside the government. If Big Sam's lips could be read, he would be silently mouthing, "don't do as I do, do as I say." Seemingly oblivious to his own internal fiscal problems, Big Sam has little hesitancy in advising Japan, Germany, and the always-in-trouble South American countries on fiscal matters, such as why they should lower or raise their interest rates.

These trusts operate in a manner similar to a chain letter. For each new person receiving benefits, it takes several new persons paying into the trusts to keep it from going bust. For Social Security, it takes about three people paying into the trust to provide benefits to one person. Thus when one new person is added to the list to receive benefits, three new persons must start paying into the trust. Hence, at some point in time a shortage of new persons will occur. And so it will about 2025. Social Security has publicly announced benefit payments will exceed receipts in 2025 as more and more baby boomers become eligible to receive benefits; indicating the number of people paying into the trust versus those receiving benefits is declining below three to one. The population explosions which were one time rampant are coming to an end—witness the declining birth rates in Japan and Italy, for example. In the U.S. it spells catastrophe.

As of the close of fiscal 2001, the Social Security trust fund had only sufficient assets, including monthly earnings on these assets, to pay benefits to current beneficiaries for 40 months. To pay benefits to these beneficiaries, until they are all deceased, funding for 229 months is required. This leaves a shortfall of 189 months which amounts to $2.8 trillion. And this is the good news. There are persons who have been paying into the trust for years, 45 or more in some cases, but as yet are not qualified to receive benefits. This liability is $11.6 trillion, bringing the total Social Security liability to $14.4 trillion. This is over twice the Public Debt. And then you have the other three trust funds, Federal Employees Retirement, Military Personnel Retirement, and Railroad Employees Retirement having a total liability of $2.4 trillion. This brings the liabilities of the four trusts to $16.8 trillion. Including the public debt, the total liability is about $23 trillion. Refer to the appendices for details on U.S. Treasury data and how these numbers are derived, all from Treasury Department statements. To put these numbers in perspective, the government's total receipts in fiscal 2001, excluding the trust funds, were only $1.0 trillion. Assuming we completely disbanded the government, other than the IRS to collect taxes, it would take over 20 years to pay off this astronomical debt. And the numbers are growing like an uncontrollable virus. **Based on Social Security estimates for 2037 of the number of persons enrolled, the total liability is some $143.3 trillion. In short, the government is severely over-extended as of now and the day of retribution is approaching.** Further, as each year passes, the problem worsens; liabilities are growing faster than receipts.

The Public Debt is a funded liability, that is, the government has issued interest bearing debt securities, T-Bills, T-Notes and Bonds, which have been sold to the public and others. Table 5.2 in Chapter 5 reveals that in fiscal 2001, 34.4 cents out of every dollar of useable receipts went to pay the interest on the Public Debt when the trust funds are excluded. This is an extravagant amount by any standard. And it has been higher. In 1991 it hit an unprecedented high of

57.9%, at which time many Fed watchers thought the Armageddon was at hand. Fortunately the good times began to roll in the late 1990s and the country was granted a reprieve, but it seems likely to be only temporary as the war on terrorism gets into full swing.

The four trust fund liabilities are referred to by the government as "unfunded liabilities," that is, underlying interest bearing debt securities have **not** been issued to cover these liabilities, otherwise we wouldn't have a problem. And why hasn't the government done this. Simply put, the government can't pay the interest expense on this astronomical debt of $16.8 trillion. **An interest expense of about 6%, which is needed to pay the benefits provided by these four trusts, would amount to around $1.0 trillion annually. This sum of money is beyond the government's capability. In fiscal 2001, the government's total receipts of $1.0 trillion from all sources, excluding the trust funds, would just cover the interest.**

Table 10.1 summarizes the liabilities of the four retirement trust funds and lists the public debt as of the close of fiscal year 2001.

The White House and Congress Don't Listen

If Congress had listened and followed the Treasury Secretary William Simon's advice given in his testimony before the House Appropriations Committee in 1976, the current trust fund problems would not be us today. In his testimony he condemned the practice of describing legislated spending for future programs as "future liabilities." He stated this was a deceptive formulation: they are present liabilities, payable in the future, and the budget should make this clear.

The Grace Commission's Report, which would have saved the government an estimated $525 billion over a three year period during the lower half of the 1980s, died without a whimper when Congress ignored it and its initiator, President Reagan, didn't push it.

David A. Stockman, who was Director of the Office of Management and Budget in the Ronald Reagan Administration, was high on Congress's hit list when he tried to pare the budget. Senator Lowell Weicker voiced his concern about Stockman when he said,

"How long are we going to allow this little pissant to dictate what we do around here? He's had his head up his ass from day one."[1]

When the book, *The Rise and Fall of the Great Powers* was released in 1987, its author, Paul Kennedy, was berated by Congress.

The White House and Congress have deaf ears; they only listen and march to their own drummer.

About This Work

This work centers on 1) the critical fiscal elements that have signaled the decline of the U.S., 2) fundamentals that will be instrumental in the approaching fall and 3) essentials that will make the demise of the U.S. apparent to its leaders and other world figures. In the near years ahead the U.S. will join England, Spain and others in the passing parade of past great nations.

This work does not presume to be an in depth scholarly analysis of **all** of the factors that have contributed to the decline and the approaching fall of the U.S. For works providing a more general and in depth analysis of the factors which have precipitated the decline and fall of other great nations, refer to: 1) *The Fate of Empires and Search for Survival*[2] by Sir John Glubb, 2) *The Rise and Decline of Nations*[3] by Mancur Olson and 3) *The Rise and Fall of the Great Powers*[4] by Paul Kennedy. For works dealing specifically with the fiscal problems of U.S. the following works are suggested: 1) *A Time for Truth*[5] by William E. Simon, former Secretary of the Treasury, 2) *The Triumph of Politics*[1] by David A. Stockman, who was Director of the Office of Management and Budget under President Reagan and 3) *How Sick is Uncle Sam?* [6] by the author of this book.

And Then There Are Other Problems

There is also another problem, the Public Debt. **In fiscal 2000, about $1.2 trillion Public Debt securities, of the some $6 trillion total, were owned by foreign and international investors; some 21.6% of the total.** Refer to Chapter 5 for information on who holds what. These investors buy U.S. debt securities because of the normally and comparatively high interest rate and for their perceived safety.

Table 10.1

Summary of Liabilities

(As of the close of fiscal 2001)

	Benefits		Total
	Present	Future	
Unfunded Liabilities			
Social Security	$ 2.8 t	$ 11.6 t	$ 14.4 t
Federal Employees Retirement -	0.8	1.7	0.9
Military Retirement	0.5	0.5	1.0
Railroad Retirement	0.1	0.4	0.5
			$ 16.8 t
Funded Liability			
Public Debt			$ 5.8
Total Liabilities			$ 22.6 t

Should either of these two parameters come into question, many foreign and international investors would move their money elsewhere. As a result, the U.S. would be forced to significantly raise the interest rate to attract new investors. Certainly not a desirable option since the government's annual interest expense would rise, compounding the U.S.'s problem.

Then there is another problem. **All projections in this work are predicated on ZERO cost of living adjustments (COLAs).** Hence, the liability projections in this work are optimistic and will undoubtedly be below actuals.

Then there is still another problem that is hard to appraise, life expectancy. As life expectancy increases, benefits will have to be paid for a longer time. To be on the low and optimistic side, current life expectancy tables were used to develop the liabilities cited herein. The progress of medical science in the next twenty years is likely to exceed public expectations, possibly wiping out dreaded cancer and some heart problems, not to mention other medical afflictions. Adding a few years of life expectancy means prolonging benefit payments—and this is big bucks, trillions. In brief, it seems highly likely the duration of benefit payments will lengthen in the future, again making the dollar numbers developed herein below actuals.

And what about the birth rate? Should this drop dramatically, as it has in Japan and Italy, it would be impossible to maintain the needed ratio of about three people paying into a trust for each person receiving benefits.

Then there is the lack of leadership. What is seemingly leadership is nothing more than the Washington contingency spouting the latest results obtained from polling the public. Around 1970 when the big IBM 360 and 370 computers became available, polling was transformed from an art to a science. This was also the time when the government's interest expense as a percent of receipts began to rise from roughly 11%, which it averaged in the 1960s, to a peak of 57.9% in 1991. The good times of the 1990s subsequently lowered the number to 34.4% in 2001, still a ridiculously high number. Turning to leadership, there is mistaken idea that the elected officials in Washington should follow the wishes of their electorate. Wrong. An official should make a judgment after considering, 1) the wishes of their electorate, 2) what is best for the nation as opposed to the electorate's section of the country, and 3) what is best for the nation with respect to world affairs. These three considerations may be widely different. Based on the official's judgment and decision, it is then up to him or her to persuade the electorate what should be done. This kind of leadership was demonstrated by President F. D. Roosevelt in the late 1930s and in 1940. Contrary to the U.S. consensus, which was to stay out of the conflict in Europe, he established the draft and the lend-lease program with England. Today the polls are the holy of holies; "I've got to get re-elected."

Probable Response to This Book

A very cool U.S. reception to this book is almost a sure thing. Many U.S. citizens will find it unpatriotic and upsetting. However, for the serious reader who takes the time to review and check the data herein, the conclusions are undeniable; the U.S. is over the hill; the fall of the U.S. is approaching.

One basic difference between this work and prior material is its specificity. Previous books lacked explicit numbers and how they were derived. Actual liability dollars of the four trusts and their modus operandi were missing. Now it is decades too late to avoid a massive restructuring of U.S. commitments and liabilities.

Unfortunately a very heady environment exists in smoggy bottom, listed on maps as Washington, D.C. The opulent offices and prestige afforded some temporary government employees, otherwise known as the President, Senators and Representatives, gives them a feeling of imperiousness. To expect anything other than ridicule and upbraiding from the Washington crowd about this work would be a bad bet. Their favorite approach for anything they didn't initiate or approve is to ignore it, hoping it will die so they can sweep it under the rug; their expertise in this area is unrivaled. Should this fail, contemptuous ridicule is likely to be used.

So what is going to happen? In the immediate future, the answer is likely "nothing." President and Congress will ignore the problem as long as possible hoping its visibility in the public eye will disappear. There will continue to be some talk about fixing Social Security by letting new participants invest elsewhere, etc., but again this shows a lack of understanding of the problem; the liabilities currently on the books won't go away. Keep in mind, for every new person receiving Social Security benefits, it takes three new persons paying into the trust. Thus with the number of beneficiaries rapidly growing as the baby boomers become eligible for benefits, the idea of letting new participants invest elsewhere is ludicrous; it would simply exacerbate the problem. Thus the Ponzi modus operandi will probably remain unchanged; benefits will continue to be paid from essentially current receipts. Social Security taxes will probably be raised from the current 15.3%; they have been raised 20 times since Social Security was created in 1937 when the tax rate was a mere 2%. Eventually the government will have to restructure its commitment to these four trust funds. By then, the numbers will be awesome, unreal and unimaginable. Again as usual, the poor will get the short end of the stick. The problem is an Armageddon because so many people are affected, probably as much as 70% of the population when all four trusts and the trickle down effect are considered. In a March 19, 2001 news release, the Acting Commissioner of Social Security stated, "Social Security is the major source of income for two-thirds of older Americans and virtually the only source of income for one-third of older Americans." Then you have to add Federal Retirees, Military Retirees and Railroad Retirees. The trickle down effect must also be added; less spending power by retirees will have a deleterious effect on business.

The damage has already been done, but the quicker it is recognized and the liabilities restructured, the lesser the damage. Social Security has reported it will run out of money in 2038 if nothing is done. At that time, the liabilities will be about $143 trillion.

The Next Leading Nation

As history has shown, there is always a nation waiting in the wings to become the leader. Further, in many instances the switch in reins occurs so gradually the populace in both countries is not aware of it until it becomes de facto.

And who is going to take the U.S.'s place. It seems reasonably clear when one looks around the world and sees who might be capable. Although not without current and future problems to overcome, it will be China. If you consider China to have been a leading nation of the world around the time of Marco Polo, and if the author's prediction comes to pass, it will be the first time in world history a nation has ever attained the leading nation status for the second time. Probably the greatest obstacle they have to overcome is their language. English is the technical language of the world, like it or not, and technology leads the parade to world power. The Chinese leaders will have to recognize this and make English a required subject in their school system. The Chinese are intelligent, industrious, and women share equal rights with men. They are one of the few nations that are making a direct and major effort to control their population. Some of their industrial plants are the most modern in the world, thanks to U.S. businessmen who, wanting to share in the Asian markets, provided the money and know-how to build them. This know-how cost U.S. businessmen years of learning and multi-billions of dollars. Essentially China got both for free. It's no secret that a large number of the products sold in the U.S. come from China, and the quality is very good. They also have atomic know-how . . . and they're not taking flak from anyone.

It's anybody's guess as to when the transition will begin to be recognized world-wide. The author's estimate is around 2050.

References:

1) Stockman, David A. *The Triumph of Politics - Why the Reagan Revolution Failed.* New York, NY: Harper & Row, Publishers, 1986: quotations with illustrations which follow p.118

2) Glubb, Sir John Bagot. *The Fate of Empires and Search for Survival* Edinburgh, Scotland: William Blackwood & Sons Ltd., 1978

3) Olson, Mancur. *The Rise and Decline of Nations.* New Haven and London: Yale University Press, 1982

4) Kennedy, Paul. *The Rise and Fall of the Great Powers.* New York: Random House, 1987

5) Simon, William E. *A Time for Truth,* New York: McGraw-Hill Book Company, 1978

6) Hadady, R. Earl. *How Sick is Uncle Sam*. Pasadena, CA: Key Books Press, 1986.

APPENDICES

The national budget must be balanced. The public debt must be reduced: the arrogance of the authorities must be moderated and controlled. Payments to foreign governments must be reduced, if Rome doesn't want to become bankrupt. People must again learn to work, instead of living on public assistance.

— *Marcus Tullius Cicero 106-43BC*
Great Roman orator, statesman,
philosopher & author

The Modus Operandi of Retirement Trust Funds

The retirement trust funds—Social Security, Federal Employees Retirement, Military Retirement, and Railroad Retirement—issue annual fiscal reports which are available to the public. The following are two excerpts from these publications.

> *Beginning in 2025, assets of the combined Old-Age and Survivors Insurance and Disability Insurance Trust Funds (Social Security) will be drawn down to pay benefits until the funds are exhausted in 2038.*
>
> — ***William A. Halter,***
> *Acting Commissioner of Social Security,*
> *News Release, March 19, 2001*

> *Cash flow problems occur only under the most pessimistic employment assumption. Even under that assumption, the cash flow problems do not occur until the year 2035.*
>
> — ***U.S. Railroad Retirement Board,***
> *Twenty-First Actuarial Valuation*

The focus of all annual reports on these four trusts are not the health of the fund, but rather, when the coffers will run dry and they will no longer be able to pay the current benefits without an infusion of new funds. Such statements are an outright admission that the modus operandi is faulty and these trusts will eventually fail. Lending credence to the Social Security problem was President Bush's announcement which appeared in Social Security Online on June 5, 2001.

President Bush Announces Social Security Commission

> President Bush has announced the establishment of a Social Security Commission on Social Security reform. The bipartisan Commission has been asked to develop a proposal to modernize and restore long-term fiscal soundness to Social Security. The commission will schedule public hearings soon and issue a final report this fall.

The operation of these four retirement trust funds is the crux of our government's eventual downfall. In the government's words, these trusts are referred to as "unfunded", vis-à-vis the public debt which is funded with debt securities. They are unfunded because the government lacks sufficient fiscal resources to issue covering debt securities and pay the accompanying interest. Consequently, benefits of these four trusts are paid on a hand-to-mouth basis. In effect current receipts must be used to pay benefits; receipts can't be held long enough to garner sufficient earnings to cover the benefit payments. This modus operandi is commonly referred to as a Ponzi scheme, which in private business has been ruled illegal by the courts. The unabridged Merriam Webster defines Ponzi as "(after Charles A. Ponzi, died 1949, Italian-born American swindler) an investment swindle in which some early investors are paid off with the money put up by later ones . . ."

Former Secretary of the Treasury, William Simon, the first legitimate doomsayer, warned Congress about this method of operation in his testimony before the House Appropriations Committee in 1976. In his testimony he condemned the practice of describing legislated spending for future programs as "future liabilities." He stated this was a deceptive formulation; that they are present liabilities, payable in the future, and the budget should make this clear. Had Congress paid attention to Secretary Simon's advice, this problem would not be with us today. But then, the White House and Congress have deaf ears and they only listen to their own drummer—now, it is the public who will be paying.

As of the end of fiscal year 2001, this seemingly innocuous but insidious operating scheme has racked up liabilities of some $16.8 trillion and it is growing significantly each year. Based on estimated data from Social Security for the year 2038, the total projected liabilities developed herein will be $143.3 trillion, see Appendix C. It makes the public debt, which is pushing $6 trillion as of the year 2001, look like pocket change. The government's total annual receipts from all sources, except the trusts, wouldn't even cover a fraction of the interest payments on such an astronomical sum.

The government's operation of these trusts is insidious. It's a catch-22. Receipts are only in the trust a fraction of the time necessary to garner the earnings needed to pay the specified benefits. For Social Security, it has been necessary to:

1. Maintain a ratio of about three persons paying into the trust to pay benefits to one person.
2. Increase taxes from time-to-time. Since the trust was first established, taxes have been raised 20 times. The actual tax has been increased from 2% to 15.3%.
3. Annually increase the maximum salary on which taxes are collected. In 1937, taxes were collected on just the first $3000 dollars of a person's earnings. By 2001, taxes were collected on the first $84,400 earned. See the Social Security Tax-rate Schedule in Appendix D.

The government's method of operating these trusts is fundamentally flawed. Like a dog chasing it tail, these trusts will eventually fail; it is only a matter of time.

How to Estimate Liabilities of Retirement Trust Funds

The fiscal condition of a given trust fund is simply the **assets** minus the **liabilities** on any specific date, such as the close of a fiscal year—much like the balance sheet of a private business. The specific date selected is referred to herein as the **cutoff date**. As of that date it is assumed that neither new funds nor new participants are accepted into the trust—the trust's capability of meeting its commitments as of that date represents its fiscal condition.

Assets

The assets are readily available from several sources; a convenient source is the *Monthly Treasury Statement of Receipts and Outlays of the U.S. Government.* Refer to Figure B-1 which is a copy of "Table 8" in an issue of that publication dated December 31, 2001, which is the end of Social Security's fiscal year 2001—vis-a'-vis the government's normal fiscal year beginning October 1st through September 30th. In Figure B-1 assets are listed under the heading Securities Held as Investments, Current Fiscal Year and in the right-most column Close of This Month. For example, the assets of Social Security (Federal old-age and survivors insurance) are listed as $1,071,795 ($ millions).

Table 8. Trust Fund Impact on Budget Results and Investment Holdings as of December 31, 2001

[$ millions]

Classification	This Month			Fiscal Year to Date			Securities held as Investments Current Fiscal Year		
							Beginning of		Close of This Month
	Receipts	Outlays	Excess	Receipts	Outlays	Excess	This Year	This Month	
Trust receipts, outlays, and investments held:									
Airport and airway	597	525	72	736	1,299	-563	13,660	13,706	13,848
Black lung disability	48	31	17	104	99	4	......	......	......
Federal disability insurance	9,652	5,305	4,347	20,731	15,639	5,092	135,842	136,536	140,947
Federal employees life and health	......	81	-81	......	-338	338	30,341	30,749	30,737
Federal employees retirement	19,738	3,998	15,740	22,902	12,099	10,803	554,346	549,402	565,129
Federal hospital insurance	22,663	11,567	11,097	47,188	35,820	11,368	197,137	197,329	208,888
Federal old-age and survivors insurance	65,121	31,663	33,458	131,868	94,203	37,665	1,034,114	1,038,508	1,071,795
Federal supplementary medical insurance	10,577	8,445	2,132	26,725	27,117	-392	41,978	39,286	40,828
Hazardous substance superfund	652	99	554	770	295	475	3,630	3,531	3,444
Highways	2,628	2,378	250	6,119	7,780	-1,661	24,115	25,306	24,955
Military advances	1,051	1,116	-65	2,397	2,738	-341	......	......	......
Military retirement	864	2,924	-2,060	25,303	8,675	16,629	156,978	175,626	173,724
Railroad retirement	-9	698	-707	778	2,091	-1,314	26,865	26,914	26,874
Unemployment	3,020	3,716	-696	7,003	10,003	-3,000	88,638	86,523	85,918
Veterans life insurance	424	25	399	459	214	246	13,462	13,299	13,695
All other trust	239	425	-186	916	1,218	-302	14,341	14,467	14,338
Total trust fund receipts and outlays and investments held from Table 6-D	**137,264**	**72,996**	**64,268**	**293,997**	**218,953**	**75,044**	**2,335,447**	**2,351,181**	**2,415,120**
Less: Interfund transactions	83,406	83,406	......	134,123	134,123	......			
Trust fund receipts and outlays on the basis of Tables 4 & 5	53,858	-10,410	64,268	159,874	84,829	75,044			
Total Federal fund receipts and outlays	**134,126**	**171,826**	**-37,700**	**306,518**	**418,647**	**-112,129**			
Less: Interfund transactions	70	70	......	83	83	......			
Federal fund receipts and outlays on the basis of Table 4 & 5	134,056	171,756	-37,700	306,436	418,565	-112,129			
Net budget receipts & outlays	**187,914**	**161,347**	**26,567**	**466,309**	**503,394**	**-37,085**			

Note: Interfund receipts and outlays are transactions between Federal funds and trust funds such as Federal payments and contributions, and interest and profits on investments in Federal securities. They have no net effect on overall budget receipts and outlays since the receipts side of such transactions is offset against bugdet outlays. In this table, Interfund receipts are shown as an adjustment to arrive at total receipts and outlays of trust funds respectively.

Figure B-1

Liabilities

The liabilities are of two types, specifically:

1. The liability to continue paying benefits to persons currently receiving benefits until they are deceased. This obligation will be referred to as **Present Benefits Liability.**
2. The liability to persons who have been paying into the trust, but as yet are not eligible by age or whatever, to receive benefits. This obligation will be referred to as **Future Benefits Liability**.

Present Benefits Liability - The method of calculating the Present Benefits Liability is straight forward and readily accomplished.

Assuming a **cutoff** date of December 31, 2001, after which neither new funds nor participants are accepted into the trust, the monthly Outlays (benefits), see Figure B-1, will decrease each month as an increasing number of beneficiaries become deceased. The average person's age when his/her benefits start and this person's life expectancy from actuary tables determine how many years benefits must be paid. Note that race and sex must be taken into account.

As a simple example, if the typical person starts receiving benefits at age 65 and actuary tables predict a life expectancy of 20 years, the benefits will decrease to 0 at the end of 20 years. Assuming the total benefits being paid were $20,000 per year at the time the analysis was initiated, the yearly benefits would decrease at the rate of $1,000 per year and at the end of 20 years would be zero. This is not a precise approach since life expectancy, cost of living, etc. adjustments will be required in the real world. However, this approach will be optimistic, it is simple, and is satisfactory for the purposes herein. The following sections itemize the necessary data to calculate the two liabilities.

The actual data required for calculating the Present Benefits Liability are:

1. The division of males, females and race at the cut-off date.
2. Average age of persons who starts receiving benefits on the cut-off date.

3. Actuarial tables to determine life expectancy of the average person who starts receiving benefits on the cut-off date—this is the number of years benefits must be paid.
4. The benefits paid in the month prior to the cut-off date.

For example, referring to Figure B-1, on the close of Social Security's fiscal year 2001, (December 31, 2001) the Treasury Department reported that Social Security had an asset balance of $1,071,795m—listed under Securities held as Investments, Current Fiscal Year, subheading Close of This Month. Remember, Social Security operates on a calendar fiscal year as opposed to the rest of the government. Turning to the benefits paid, they are listed under This Month, subheading Outlays. For December 2001, the last month of fiscal year 2001, the Outlays were $31,663m. The actuary tables used predicted the average new beneficiary aged 65, who begins receiving benefits in January 2002, has a life expectancy of 19.1 years (229.2 months).

Based on data in the preceding paragraph, benefits will decrease from $31,663m by $31,663/229.2 each month and becomes zero at the end of 229.2 months. On the other hand, the assets will earn interest each month—assumed to be 0.5% monthly, slightly over 6% annually. Thus at the end of January 2002, the asset balance in the account in millions will be:

$$\$1{,}071{,}795 - (31{,}663 - 31{,}663/229.2) + (1{,}071{,}795 \times 0.005) =$$
$$\$1{,}071{,}795 - 31{,}525 + 5{,}359 = \$1{,}045{,}629\text{m}.$$

Using a computer and setting up a spread sheet, it is easy to determine 1) the asset balance each month, 2) the number of months before the assets are exhausted, and 3) the assets needed to continue to pay benefits until all recipients are deceased..

A separate appendix covers the present liability calculations for each trust. The data elements vary from trust to trust and are specified in the introduction of each appendix.

Future Benefits Liability – The data needed to determine this liability are:

1. Annual trust fund receipts for past years
2. Number of years the average participant has paid into trust before receiving benefits.

The underlying principles behind determining Future Benefits is somewhat complex but determining this number is quite simple.

The means of approximately determining this liability is readily seen by referring to the graph in Figure B-2, the parameters of which are applicable to Social Security. For Social Security, it is assumed that the typical person pays into the trust for 45 years, that is, begins work and starts paying into the trust at age 20 and retires at 65. However, changing the parameters, such as the average age at retirement, would make it applicable to other trusts. The series of horizontal lines represent age groups separated by 5 years. Lines representing age groups every year, instead of 5, would reveal the same information, but the chart would become unnecessarily busy.

It is clear that going backward in time from the present date of 2000 on the graph, less and less of the past trust receipts represent a liability to pay benefits in the future. For the most recent year on the graph, the year 2000, all of the receipts represent a liability to pay benefits in the future. Since the involved relationships are linear and a time span of 45 years is involved for Social Security, only 1/45 of the payments into the trust 45 years ago plus earnings are a liability to pay future benefits; those particular participants will be receiving benefits next year. Similarly, 2/45th of the receipts 44 years ago plus earnings is a liability to pay future benefits, and so on and on until 45/45th of the current receipts is an obligation.

In reviewing the chart in Figure B-2, it may at first seem that Future Benefit Liabilities could be determined by simply taking half of the total funds collected during the period in question. This would represent the total money put into the trust by an individual, but it would not include the annual compounded earnings—a significant amount when 45 years are involved.

To provide a growth perspective, Social Security liabilities are included in Appendix C for the years 1981, 1989, 2001 and 2037. The rate of growth over these years is awesome; an indication of the approaching fall.

The **Future Benefits Liability** is the biggie in Social Security; the **Present Benefits Liability** is only the tip of the iceberg.

Social Security Future Benefit Liabilities versus Prior Receipts

This graph illustrates how to approximate a trust's liability to participants who have been paying into it for years but as yet are not eligible, by age etc., to receive benefits. In this illustration the year is 2000. The various horizontal lines represent participants differing 5 years in age; the dotted part of the line, the age from 0 to 20. Participants start paying into the trust at age 20; they start receiving benefits at age 65. Note that the trust liability is greatest for the most recent year and decreases linearly to the date when the first payment into the trust was made.

Example: The bottom line in the shaded section represents participants born in 1935, who started work in 1955 at age 20, paid into the trust through 1999 when they reached the age of 65 and are now eligible to receive benefits beginning in the year 2000.

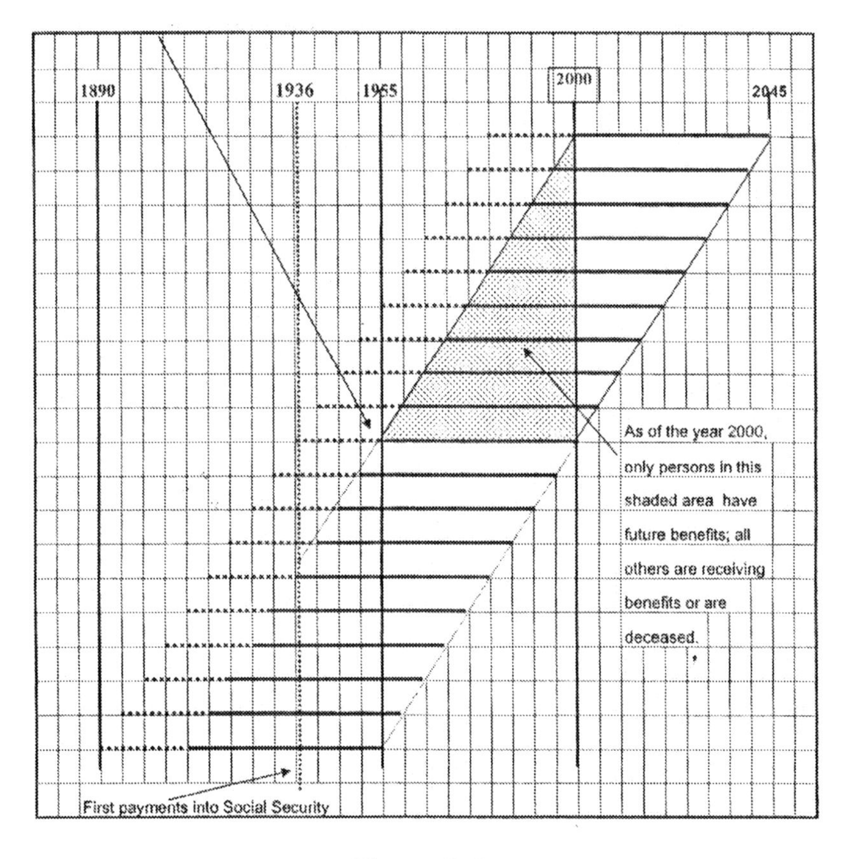

Figure B-2

Social Security Trust Liabilities

In order to provide an overall perspective, this Appendix covers Social Security Liabilities for past years 1981, 1989, and 2001 and the future year 2038. Social Security works on a calendar year as opposed to the normal government fiscal year which begins on October 1st and extends through September 30th.

At any specified time, such as the end of a fiscal year, Social Security has these two basic liabilities:

1) **Present Benefits Liability** – the obligation to pay benefits to current beneficiaries until they are deceased
2) **Future Benefits Liability** – the obligation to pay benefits to covered workers—a Social Security phrase meaning a person who is paying into Social Security but as yet is not qualified to receive benefits. At the specified time it is the equivalent of what the worker has paid into the trust plus reasonable earnings.

See Appendix B for how these liabilities are calculated.

The following is a summary of Social Security liabilities at the close of various fiscal years:

Calendar Year	**Benefit Liabilities**		
	Present ($s trillions)	Future ($s trillions)	**Total** ($s trillions)
1981	1.208 t	$ 1.627 t	$ 2.835 t
1989	1.862	3.989	5.851
2001	2.841	11.601	14.442
2038	41.141	106.077	147.218

Social Security *Present* Benefits Liability as of 1981

The following Table C.1 reveals assets are only sufficient of paying benefits for 2 months before they are depleted. This results in a **Present Benefits Liability** of **$1.208 trillion,** as of the close of calendar year 1981—the shortage of funds needed to pay beneficiaries until they are deceased

The **Future Benefits Liability** follows this presentation.

The **Present Benefits Liability** spread sheet assumes:

1) Beneficiaries are equally divided between the sexes.
2) Participants retire at age 65 and die at 84.1; an average life expectancy at retirement of 19.1 years (229.2 months)—data are from actuary tables.
3) The total monthly benefit payments linearly declines and reaches zero in 19.1 years as beneficiaries become deceased.
4) Cost of living (COLAs) and other adjustments, which would have increased the liabilities, have been ignored
5) Assets earn interest at the rate of 0.5% each month; slightly over 6% per year. Earnings are added to assets each month before benefit payments are deducted.
6) In the spread sheet, the first entries under Assets and Benefit Payments are from the September 30th, 1981 issue of the *Monthly Treasury Statement of Receipts and Outlays*

Table C.1

Year Calendar	**Month**	**Assets*** Beginning of Month ($s m) (1)	End of Month ($s m) (2)	**Earnings** At 0.5% per month ($s m) (3)	**Benefit Payments** (end of month) ($s m) (4)
1981	12		23,255		10,775
1982	1	23,255	12,643	116	10,728
	2	12,643	2,026	63	10,681
	3	2,026	-8,598	10	10,634
	4		Assets exhausted		10,587
	5				10,540
	6				10,493
	7				10,446
	8				10,399
	9				10,352
	10				10,305
	11				10,258
	12				10,211
1983	13				10,164
	14				10,117
	15				10,070
....					
	218				527
	219				480
	220				433
	221				385
	222				338
	223				291
	224				244
	225				197
	226				150
	227				103
	228				56
2001	229				9
	230				-38
		Sum of months 3 through 229 **Present Benefits Liability as of 1981**			1,208,022

* Calculation basis is: Column (1) + Column (3) - Column (4) = Column (2)

.... Month rows 16 through 217 compressed and hidden.

Social Security *Future* Benefits Liability as of 1981

The Table C.2 which follows reveals a **Future Benefits Liability** of **$1.627 trillion**, as of the close of calendar year 1981—the shortage of funds needed to pay participants who will become beneficiaries in the future.

Social Security operates on a calendar year basis versus the federal government's fiscal year which begins October 1st and ends September 30th.

For the **Present Benefits Liability** see the preceding pages.

In the spread sheet which follows, calculation of the **Future Benefits Liability** is based on:

1) the procedure and graph in Appendix B.
2) participation in the trust beginning at age 20, retiring at 65.
3) data furnished by Social Security.

Table C.2

Year	Receipts		Liability			
			Fractional		Annual	Cumulative
(calendar)	$s m	Growth	Part	$s m	$s m*	$s m**
1937	767		1/45	17	18	18
1938	375	-51.1%	2/45	17	18	37
1939	607	61.9%	1/15	40	43	82
1940	368	-39.4%	4/45	33	35	122
1941	845	129.6%	1/9	94	100	228
1942	1,085	28.4%	2/15	145	153	395
1943	1,328	22.4%	7/45	207	219	638
1944	1,422	7.1%	8/45	253	268	944
1945	1,420	-0.1%	1/5	284	301	1,302
1946	1,447	1.9%	2/9	322	341	1,721
1947	1,722	19.0%	11/45	421	446	2,270
1948	1,969	14.3%	4/15	525	557	2,963
1949	1,816	-7.8%	13/45	525	556	3,697
1950	2,928	61.2%	14/45	911	966	4,885
1951	3,784	29.2%	1/3	1,261	1,337	6,515
1952	4,184	10.6%	16/45	1,488	1,577	8,482
•••••••						
1966	23,381	30.9%	2/3	15,587	16,523	146,076
1967	26,413	13.0%	31/45	18,196	19,287	174,128
1968	28,493	7.9%	32/45	20,262	21,477	206,053
1969	33,346	17.0%	11/15	24,454	25,921	244,337
1970	36,993	10.9%	34/45	27,950	29,627	288,624
1971	40,908	10.6%	7/9	31,817	33,726	339,668
1972	45,622	11.5%	4/5	36,498	38,687	398,736
1973	54,787	20.1%	37/45	45,047	47,750	470,410
1974	62,066	13.3%	38/45	52,411	55,556	554,190
1975	67,640	9.0%	13/15	58,621	62,139	649,580
1976	75,034	10.9%	8/9	66,697	70,699	759,254
1977	81,982	9.3%	41/45	74,695	79,176	883,986
1978	91,903	12.1%	14/15	85,776	90,923	1,027,947
1979	105,864	15.2%	43/45	101,159	107,228	1,196,853
1980	119,712	13.1%	44/45	117,052	124,075	1,392,739
1981	142,438	19.0%	1	142,438	150,984	1,627,287

Future Benefits Liability as of 1981 (→ 1,627,287)

••••••• Year rows 1955 through 1965 compressed and hidden

* Fractional dollars + annual earnings @ 6%

** Prior year + annual earnings @ 6% + current year

Social Security *Present* Benefits Liability as of 1989

The following Table C.3 reveals assets are only sufficient of paying benefits for 8 months before they are depleted. This results in a **Present Benefits Liability** of **$1.862 trillion**, as of the close of calendar year 1989—the shortage of funds needed to pay beneficiaries until they are deceased

The **Future Benefits Liability** follows this presentation.

The **Present Benefits Liability** spread sheet assumes:

1) Beneficiaries are equally divided between the sexes.
2) Participants retire at age 65 and die at 84.1; an average life expectancy at retirement of 19.1 years (229.2 months)—data are from actuary tables.
3) The total monthly benefit payments linearly declines and reaches zero in 19.1 years as beneficiaries become deceased.
4) Cost of living (COLAs) and other adjustments, which would have increased the liabilities, have been ignored
5) Assets earn interest at the rate of 0.5% each month; slightly over 6% per year. Earnings are added to assets each month before benefit payments are deducted.
6) In the spread sheet, the first entries under Assets and Benefit Payments are from the September 30th, 1989 issue of the *Monthly Treasury Statement of Receipts and Outlays.*

Table C.3

Year Calendar	**Month**	**Assets*** Beginning of Month ($s m) (1)	End of Month ($s m) (2)	**Earnings** At 0.5% per month ($s m) (3)	**Benefit Payments** (end of month) ($s m) (4)
1989	12		148,565		17,528
1990	1	148,565	131,856	743	17,452
	2	131,856	115,141	659	17,375
	3	115,141	98,418	576	17,299
	4	98,418	81,688	492	17,222
	5	81,688	64,950	408	17,146
	6	64,950	48,206	325	17,069
	7	48,206	31,454	241	16,993
	8	31,454	14,695	157	16,916
	9	14,695	-2,071	73	16,840
	10	-2,071	-18,844	-10	16,763
	11				16,687
....					
2008	217				933
	218				857
	219	Asset exhausted			780
	220				704
	221				627
	222				551
	223				474
	224				398
	225				321
	226				245
	227				168
	228				92
2009	229				15
	230				-61

Sum of months 9 through 229 → 1,862,480
Present Benefits Liability as of 1989

* Calculation basis is: Column (1) + Column (3) - Column (4) = Column (2)

.... Months rows 12 through 216 compressed and hidden

Social Security *Future* Benefits Liability as of 1989

The Table C.4 which follows reveals a **Future Benefits Liability** of **$3.989 trillion**, as of the close of calendar year 1989—the shortage of funds needed to pay participants who will become beneficiaries in the future.

Social Security operates on a calendar year basis versus the federal government's fiscal year which begins October 1st and ends September 30th.

For the **Present Benefits Liability** see the preceding pages.

In the spread sheet which follows, calculation of the **Future Benefits Liability** is based on:

1) the procedure and graph in Appendix B.
2) participation in the trust beginning at age 20, retiring at 65.
3) data furnished by Social Security.

Table C.4

Year	Receipts		Liability			
			Fractional		Annual	Cumulative
(calendar)	$s m	Growth	Part	$s m	$s m*	$s m**
1945	1,420		1/45	32	33	33
1946	1,447	1.9%	2/45	64	68	104
1947	1,722	19.0%	1/15	115	122	232
1948	1,969	14.3%	4/45	175	186	431
1949	1,816	-7.8%	1/9	202	214	671
1950	2,928	61.2%	2/15	390	414	1,125
1951	3,784	29.2%	7/45	589	624	1,816
1952	4,184	10.6%	8/45	744	788	2,714
1953	4,359	4.2%	1/5	872	924	3,801
1954	5,610	28.7%	2/9	1,247	1,321	5,350
1955	6,167	9.9%	11/45	1,507	1,598	7,269
1956	6,697	8.6%	4/15	1,786	1,893	9,598
1957	8,090	20.8%	13/45	2,337	2,477	12,651
1958	9,108	12.6%	14/45	2,834	3,004	16,414
1959	9,516	4.5%	1/3	3,172	3,362	20,761
1960	12,445	30.8%	16/45	4,425	4,690	26,697
1961	12,937	4.0%	17/45	4,887	5,181	33,480
1962	13,699	5.9%	2/5	5,480	5,808	41,297
1976	75,034	10.9%	32/45	53,358	56,559	550,983
1977	81,982	9.3%	11/15	60,120	63,727	647,770
1978	91,903	12.1%	34/45	69,438	73,604	760,240
1979	105,864	15.2%	7/9	82,339	87,279	893,133
1980	119,712	13.1%	4/5	95,770	101,516	1,048,237
1981	142,438	19.0%	37/45	117,116	124,143	1,235,274
1982	147,913	3.8%	38/45	124,904	132,399	1,441,789
1983	171,266	15.8%	13/15	148,431	157,336	1,685,633
1984	186,637	9.0%	8/9	165,900	175,854	1,962,624
1985	203,540	9.1%	41/45	185,448	196,574	2,276,956
1986	216,833	6.5%	14/15	202,377	214,520	2,628,093
1987	231,039	6.6%	43/45	220,771	234,017	3,019,796
1988	263,469	14.0%	44/45	257,614	273,071	3,474,055
1989	289,448	9.9%	1	289,448	306,815	3,989,313

Future Benefits Liability as of 1989

······ Year rows 1963 through 1975 compressed and hidden

* Fractional dollars + annual earnings @ 6%

** Prior year + annual earnings @ 6% + current year

Social Security *Present* Benefits Liability as of 2001

The following Table C.5 reveals assets are only sufficient of paying benefits for 40 months before they are depleted. This results in a **Present Benefits Liability** of **$2.841 trillion**, as of the close of calendar year 2001—the shortage of funds needed to pay beneficiaries until they are deceased

The **Future Benefits Liability** follows this presentation.

The **Present Benefits Liability** spread sheet assumes:

1) Beneficiaries are equally divided between the sexes.
2) Participants retire at age 65 and die at 84.1; an average life expectancy at retirement of 19.1 years (229.2 months)—data are from actuary tables.
3) The total monthly benefit payments linearly declines and reaches zero in 19.1 years as beneficiaries become deceased.
4) Cost of living (COLAs) and other adjustments, which would have increased the liabilities, have been ignored
5) Assets earn interest at the rate of 0.5% each month; slightly over 6% per year. Earnings are added to assets each month before benefit payments are deducted.
6) In the spread sheet, the first entries under Assets and Benefit Payments are from the September 30th, 2001 issue of the *Monthly Treasury Statement of Receipts and Outlays.*

Table C.5

Year Calendar	**Month**	**Assets*** Beginning of Month ($s m) (1)	End of Month ($s m) (2)	**Earnings** At 0.5% per month ($s m) (3)	**Benefit Payments** (end of month) ($s m) (4)
2001	12		1,212,533		36,576
2002	1	1,212,533	1,182,179	6,063	36,416
	2	1,182,179	1,151,833	5,911	36,257
	3	1,151,833	1,121,495	5,759	36,097
	4	1,121,495	1,091,165	5,607	35,938
	5	1,091,165	1,060,843	5,456	35,778
	6	1,060,843	1,030,528	5,304	35,619
	7	1,030,528	1,000,222	5,153	35,459
	8	1,000,222	969,924	5,001	35,299
	9	969,924	939,634	4,850	35,140
	10	939,634	909,352	4,698	34,980
	11	909,352	879,078	4,547	34,821
	12	879,078	848,812	4,395	34,661
2003	13	848,812	818,555	4,244	34,501
	35	185,131	155,066	926	30,991
	36	155,066	125,010	775	30,831
2005	37	125,010	94,964	625	30,671
	38	94,964	64,927	475	30,512
	39	64,927	34,899	325	30,352
	40	34,899	4,881	174	30,193
	41	4,881	-25,128	24	30,033
	42	-25,128	-55,127	-126	29,874
	43				29,714
	225				670
	226				511
	227				351
	228				191
2021	229				32
	230				-128

Assets exhausted

Sum of months 41 through 229 → 2,841,151
Present Benefits Liability as of 2001

* Calculation basis is: Column (1) + Column (3) - Column (4) = Column (

.... Month rows 14-34 and 44-224 compressed and hidden

Social Security *Future* Benefits Liability as of 2001

The Table C.6 which follows reveals a **Future Benefits Liability** of **$11.602 trillion**, as of the close of calendar year 2001—the shortage of funds needed to pay participants who will become beneficiaries in the future.

Social Security operates on a calendar year basis versus the federal government's fiscal year which begins October 1st and ends September 30th.

For the **Present Benefits Liability** see the preceding pages.

In the spread sheet which follows, calculation of the **Future Benefits Liability** is based on:

1) the procedure and graph in Appendix B.
2) participation in the trust beginning at age 20, retiring at 65.
3) data furnished by Social Security.

Table C.6

Year	Receipts		Liability			
			Fractional		Annual	Cumulative
(calendar)	$s m	Growth	Part	$s m	$s m*	$s m**
1957	8,090		1/45	180	191	191
1958	9,108	12.6%	2/45	405	429	631
1959	9,516	4.5%	1/15	634	672	1,341
1960	12,445	30.8%	4/45	1,106	1,173	2,594
1961	12,937	4.0%	1/9	1,437	1,524	4,274
1962	13,699	5.9%	2/15	1,827	1,936	6,466
1963	16,227	18.5%	7/45	2,524	2,676	9,530
1964	17,476	7.7%	8/45	3,107	3,293	13,395
1965	17,857	2.2%	1/5	3,571	3,786	17,984
1966	23,381	30.9%	2/9	5,196	5,508	24,571
1967	26,413	13.0%	11/45	6,457	6,844	32,889
1968	28,493	7.9%	4/15	7,598	8,054	42,917
1969	33,346	17.0%	13/45	9,633	10,211	55,703
1970	36,993	10.9%	14/45	11,509	12,199	71,245
......						
1986	216,833	6.5%	2/3	144,555	153,229	1,596,958
1987	231,039	6.6%	31/45	159,160	168,710	1,861,485
1988	263,469	14.0%	32/45	187,356	198,597	2,171,772
1989	289,448	9.9%	11/15	212,262	224,998	2,527,075
1990	315,443	9.0%	34/45	238,335	252,635	2,931,335
1991	329,676	4.5%	7/9	256,415	271,800	3,379,014
1992	342,591	3.9%	4/5	274,073	290,517	3,872,272
1993	355,578	3.8%	37/45	292,364	309,906	4,414,515
1994	381,111	7.2%	38/45	321,827	341,137	5,020,522
1995	399,497	4.8%	13/15	346,231	367,005	5,688,758
1996	424,451	6.2%	8/9	377,290	399,927	6,430,011
1997	457,668	7.8%	41/45	416,986	442,006	7,257,817
1998	489,204	6.9%	14/15	456,590	483,986	8,177,272
1999	526,582	7.6%	43/45	503,178	533,369	9,201,277
2000	568,433	7.9%	44/45	555,801	589,149	10,342,503
2001	602,003	5.9%	1	602,003	638,123	11,601,177

Future Benefits Liability as of 2001

...... Year rows 1971 through 1985 compressed and hidden

* Fractional dollars + annual earnings @ 6%

** Prior year + annual earnings @ 6% + current year

Table C.7

Social Security Data

The data on this page and the following three pages has been provided by Social Security to enable projections for the year 2038.

Covered Workers per Beneficiary - *Historical*

Calendar Year	Covered Workers (thousands)	Beneficiaries OASDI (thousands)	Covered Workers per Beneficiary
Historical Data			
1945	46,390	1,106	41.9
1950	48,280	2,930	16.5
1955	65,200	7,563	8.6
1960	72,530	14,262	5.1
1965	80,680	20,157	4.0
1970	93,090	25,168	3.7
1975	100,200	31,125	3.2
1980	113,649	35,118	3.2
1985	120,565	36,650	3.3
1990	133,672	39,470	3.4
1991	132,969	40,172	3.3
1992	133,890	41,029	3.3
1993	136,117	41,840	3.3
1994	138,192	42,516	3.3
1995	141,027	43,108	3.3
1996	143,505	43,498	3.3
1997	146,305	43,793	3.3
1998	149,096	44,076	3.4
1999	151,186	44,367	3.4
2000	152,903	45,166	3.4

Data Source: Social Security 2001 Trustee Report

Notes: 1) Covered Worker is one who is paying into Social Security but as yet is not qualified to receive benefits.

2) Beneficiary is one currently receiving benefits.

3) OASDI is Old Age Social Security and Disability Insurance.

Table C.8

Covered Workers per Beneficiary - *Projections*

Calendar Year	Covered Workers (thousands)	Beneficiaries OASDI (thousands)	Covered Workers per Beneficiary
Low Projection			
2005	160,444	47,558	3.4
2010	167,454	51,547	3.2
2015	172,934	58,012	3.0
2020	176,716	65,566	2.7
2025	180,036	72,738	2.5
2030	183,989	78,020	2.4
2035	189,165	81,018	2.3
2040	195,085	81,971	2.4
2045	201,290	83,073	2.4
2050	207,446	84,806	2.4
2055	213,805	87,539	2.4
2060	220,579	90,474	2.4
2065	227,831	93,269	2.4
2070	235,318	96,067	2.4
2075	242,847	99,159	2.4
Intermediate Projection			
2005	158,653	48,036	3.3
2010	164,125	52,760	3.1
2015	168,461	60,068	2.8
2020	171,234	68,474	2.5
2025	173,314	76,415	2.3
2030	175,562	82,495	2.1
2035	178,416	86,321	2.1
2040	181,385	88,036	2.1
2045	184,071	89,702	2.1
2050	186,389	91,738	2.0
2055	188,507	94,596	2.0
2060	190,555	97,604	2.0
2065	192,595	100,508	1.9
2070	194,551	103,252	1.9
2075	196,377	105,925	1.9

Data Source: Social Security 2001 Trustee Report

Notes: 1) Covered Worker is one who is paying into Social Security but as yet is not qualified to receive benefits.
2) Beneficiary is one currently receiving benefits.
3) OASDI is Old Age Social Security and Disability Insurance.

Table C.9

Asset Projections

($s in millions)

Calendar Year	Income: Receipts $s	Income: Interest $s	Income: Total $s	Outlays $s	Assets Year End $s
Low Cost Projection					
2001	534.2	73.0	607.3	437.7	1,219.0
2002	564.9	83.2	648.1	457.6	1,409.6
2003	593.4	94.0	687.4	478.4	1,618.5
2004	623.1	105.6	728.6	499.7	1,847.5
2005	654.5	118.9	773.4	523.0	2,097.8
2006	686.0	133.4	819.4	547.8	2,369.3
2007	719.8	149.2	869.0	574.9	2,663.4
2008	753.3	166.5	919.8	604.9	2,978.4
2009	789.1	185.4	974.4	639.0	3,313.8
2010	826.1	205.6	1,031.7	675.9	3,669.6
2015	1,033.9	324.4	1,358.4	931.1	5,688.0
2020	1,279.2	453.8	1,733.0	1,297.3	7,875.0
2025	1,580.5	581.6	2,162.0	1,738.1	10,024.2
2030	1,956.7	705.5	2,662.3	2,244.9	12,109.8
2035	2,433.2	834.5	3,267.8	2,799.6	14,308.0
2040	3,029.0	994.8	4,023.8	3,395.2	17,087.6
2045	3,768.7	1,214.6	4,983.3	4,122.0	20,909.4
2050	4,683.7	1,510.2	6,193.9	5,054.9	26,031.0
2055	5,821.9	1,892.2	7,714.1	6,271.7	32,623.9
2060	7,244.1	2,373.1	9,617.2	7,793.0	40,923.3
2065	9,022.2	2,987.4	12,009.6	9,659.8	51,547.5
2070	11,236.4	3,784.3	15,020.7	11,964.7	65,341.9
2075	13,982.6	4,816.4	18,799.0	14,864.5	83,192.7

Data Source: Social Security

Table C.10

Asset Projections

($s in millions)

Calendar Year	Income Receipts $s	Income Interest $s	Income Total $s	Outlays $s	Assets Year End $s
Intermediate Cost Projection					
2000	531.6	72.7	604.3	438.9	1,214.9
2001	560.1	82.1	642.3	459.9	1,397.2
2002	588.5	92.8	681.3	483.7	1,594.8
2003	617.9	104.5	722.4	510.2	1,807.0
2004	650.0	117.7	767.7	539.6	2,035.2
2005	682.3	132.1	814.4	571.5	2,278.1
2006	717.4	147.3	864.7	606.8	2,536.1
2007	752.7	163.6	916.3	645.9	2,806.5
2008	790.9	180.5	971.5	690.0	3,087.9
2009	830.7	198.1	1,028.8	737.8	3,378.9
2015	1,059.0	295.1	1,354.1	1,057.8	4,887.9
2020	1,336.2	374.3	1,710.5	1,518.4	6,104.8
2025	1,683.2	405.3	2,088.5	2,102.6	6,491.4
2030	2,120.5	354.9	2,475.4	2,807.7	5,507.5
*2035	2,675.7	188.6	2,864.3	3,624.2	2,599.1
High Cost Projection					
2001	520.6	71.8	592.4	440.5	1,201.4
2002	538.2	78.9	617.1	464.9	1,359.6
2003	575.8	92.6	668.4	493.6	1,528.3
2004	600.4	108.3	708.7	593.1	1,703.9
2005	630.1	119.0	749.1	581.5	1,871.5
2006	673.6	131.3	804.9	621.2	2,055.2
2007	715.2	143.8	859.0	663.4	2,250.7
2008	755.5	156.2	911.7	711.1	2,451.4
2009	797.9	168.4	966.6	766.1	2,651.9
2010	842.6	181.2	1,023.8	825.6	2,850.1
2015	1,097.0	229.5	1,326.5	1,219.8	3,629.3
2020	1,413.6	228.2	1,641.8	1,792.2	3,470.6
*2025	1,818.5	110.1	1,928.6	2,564.2	1,367.2

*Estimates for later years are not shown because the combined OASI and DI Trust Funds are estimated to become exhausted in 2038 under the Intermediate Cost assumption—in 2027 under the High Cost assumption.

Note: Totals do not necessarily equal the sums of rounded components

Data Source: Social Security

Social Security Present Benefits Liability as of 2038

This liability estimate is based on Social Security's Intermediate Cost Projection which forecasts the exhaustion all of their assets by the year 2038—see Table C.10.

The **Present Benefits Liability** in Table C.11 is estimated to be **$41.141 trillion**, as of the close of calendar and fiscal year 2038—the shortage of funds needed to pay beneficiaries until they are deceased

The **Future Benefits Liability** follows this presentation.

The **Present Benefits Liability** spread sheet assumes::

1) Beneficiaries are equally divided between the sexes.
2) Participants retire at age 65 and die at 84.1; an average life expectancy at retirement of 19.1 years (229.2 months)—data are from actuary tables.
3) The total monthly benefit payments linearly declines and reaches zero in 19.1 years as beneficiaries become deceased.
4) Cost of living (COLAs) and other adjustments, which would have increased the liabilities, have been ignored
5) Assets earn interest at the rate of 0.5% each month; slightly over 6% per year. Earnings are added to assets each month before benefit payments are deducted.
6) In the spread sheet, the first entries under Assets and Benefit Payments are from estimates provided by Social Security—see Table C.10.

Table C.11

Year Calendar	**Years** Benefits Provided	**Assets** Beginning of Year ($ billions)	End of Year ($ billions)	**Earnings** At 6% per year ($ billions)	**Benefit Payments** (end of year) ($ billions)
2030					2,807.7 (1)
2035					3,624.2 (1)
2,036					3,787.5 (2)
2,037		exhausted			3,950.8 (2)
2,038	1	0	0	0	4,114.1 (2)
2,039	2	0	0	0	3,897.6
2,040	3	0	0	0	3,681.0
2,041	4	0	0	0	3,464.5
2,042	5	0	0	0	3,248.0
2,043	6	0	0	0	3,031.4
2,044	7	0	0	0	2,814.9
2,045	8	0	0	0	2,598.4
2,046	9	0	0	0	2,381.8
2,047	10	0	0	0	2,165.3
2,048	11	0	0	0	1,948.8
2,049	12	0	0	0	1,732.3
2,050	13	0	0	0	1,515.7
2,051	14	0	0	0	1,299.2
2,052	15	0	0	0	1,082.7
2,053	16	0	0	0	866.1
2,054	17	0	0	0	649.6
2,055	18	0	0	0	433.1
2,056	19	0	0	0	216.5
2,057		0	0	0	0.0
2,058					
		Sum of years 1 through 19			**41,141.0**

Present Benefits Liability as of 2038 →

Note: $s are in billions versus millions in prior tables

(1) Social Security Intermediate Asset projections - see Table C.10

(2) Annual Benefit Payment estimates based on Social Security Intermediate Projection

Social Security *Future* Benefits Liability as of 2038

Table C.12 which follows reveals a **Future Benefits Liability** of **$106.077trillion**, as of the close of calendar year 2038—the shortage of funds needed to pay participants who will become beneficiaries in the future.

Social Security operates on a calendar year basis versus the federal government's fiscal year which begins October 1st and ends September 30th.

For the **Present Benefits Liability** see the preceding pages.

In the spread sheet which follows, calculation of the **Future Benefits Liability** is based on:

1) the procedure and graph in Appendix B.
2) participation in the trust beginning at age 20, retiring at 65.
3) data furnished by Social Security.
5) dollars are in billions vis-à-vis prior spread sheets in millions.

Table C.12

Year	Receipts		Liability			
			Fractional		Annual	Cumulative
(calendar)	$s b	Growth	Part	$s b	$s b*	$s b**
1994	381		1/45	8	9	9
1995	400	4.8%	2/45	18	19	28
1996	425	6.3%	1/15	28	30	60
1997	458	7.8%	4/45	41	43	107
1998	489	6.9%	1/9	54	58	171
1999	527	7.6%	2/15	70	74	255
2000	568	7.9%	7/45	88	94	365
2001	604	6.3%	8/45	107	114	500
2002	642	6.3%	1/5	128	136	666
2003	681	6.1%	2/9	151	160	867
2004	722	6.0%	11/45	177	187	1,106
2018	1,553	4.9%	5/9	863	914	12,329
2019	1,629	4.9%	26/45	941	998	14,066
2020	1,711	5.0%	3/5	1,026	1,088	15,998
2021	1,779	4.0%	28/45	1,107	1,173	18,131
2022	1,850	4.0%	29/45	1,192	1,264	20,483
2023	1,924	4.0%	2/3	1,283	1,360	23,071
2024	2,001	4.0%	31/45	1,378	1,461	25,917
2025	2,089	4.4%	32/45	1,485	1,574	29,046
2026	2,162	3.5%	11/15	1,585	1,680	32,469
2027	2,237	3.5%	34/45	1,690	1,792	36,209
2028	2,316	3.5%	7/9	1,801	1,909	40,291
2029	2,397	3.5%	4/5	1,917	2,032	44,741
2030	2,475	3.3%	37/45	2,035	2,157	49,582
2031	2,550	3.0%	38/45	2,153	2,282	54,840
2032	2,626	3.0%	13/15	2,276	2,413	60,543
2033	2,705	3.0%	8/9	2,404	2,549	66,724
2034	2,786	3.0%	41/45	2,538	2,691	73,418
2035	2,864	2.8%	14/15	2,673	2,834	80,657
2036	2,950	3.0%	43/45	2,819	2,988	88,484
2037	3,039	3.0%	44/45	2,971	3,149	96,943
2038	3,130	3.0%	1	3,130	3,318	106,077

Future Benefits Liability as of 2038 $106.1 trillion

****** Year rows 2005 through 2017 compressed and hidden

* Fractional dollars + annual earnings @ 6%

** Prior year + annual earnings @ 6% + current year

Social Security Benefits No Windfall

There are some people, including Bill Archer, former U.S. Republican Representative from Texas, that believe the typical Social Security recipient receives much more than they ever contributed to the fund.

> *"Many of today's retirees don't realize how fast they get their money back—plus interest—once they start collecting benefits. An average earner who retires this year at 65 will get back all his or her contributions plus interest in eight years. For workers who always earned the maximum the payback time is abut 11 years—still a very good return given that at 65 the average man is now expected to live past age 80, the average woman to almost 85."*
>
> — ***Bill Archer (R-Texas)***[1]
> *U.S. Representative*

There were a couple of things that Congressman Archer failed to mention. First and most importantly, he apparently only included half of the tax that was paid into Social Security for the retiree—only what a retired employee paid. The tax contributed by the employer for the employee was left out. Possibly he thinks the money contributed by the employer belongs to the government. How about the self-employed persons who pays both the employee's and the employer's tax? No mention of this. Since 1984, a self-employed person has been paying double the amount that an employed person pays. Perhaps he thinks that the Social Security program should differ from pension plans in private industry. A private company's contribution is always considered as belonging to the employee once the employee has worked a prerequisite number of years. After the prescribed period, the employee is said to have a fully vested interest in what the employer had contributed. When you double the number of years, from 8 to 16 and 11 to 22, needed to get one's money back as specified by Congressman Archer, where is this bargain he talks about? The recipient may have become deceased before then. A second point, Congressman Archer did not specify the interest rate, how it was applied, and over what period.

In short, it is wise to question any and all similar statements unless it is backed with specifics and data.

Back to the question of how much money does one get back from Social Security, it is impossible to categorically answer this question because the Social Security rolls include some 30% disabled persons, widows, and surviving children, Only a study by the Social Security Administration can provide an accurate answer.

However, the information and data included in an April 15, 1996 article "No Bonanza", written by the author for *Barron's*, covers a hypothetical situation which reveals that Social Security is no bargain. The basic information in that article follows.

John Doe was born in January 1930. In January 1950, at age 20, Doe began work at a starting salary of $500 per month. John received annual salary increases of 4% during his career and worked until retirement in January 1995, when he reached the age of 65. Based on the Social Security tax rate schedule shown in Table D.1, the payments by Doe and his employer are tabulated in Table D.2. The total payments into Social Security amount to $64,224 during the 45 year period.

If John Doe had been able to put the same tax money into a retirement pool investing in 52 week T-Bills, his contributions would have grown to $170,169, also refer to Table D.2.

For simplicity, it was assumed that Doe's contributions were made annually rather than quarterly, as is the case with Social Security. Further, it was assumed that earnings on his contributions were not credited to his account until the end of the following year. In actual practice, Doe's equity in the pool would be somewhat greater using quarterly contributions and compounding earnings. However, no deduction was made for the cost of administering the pool.

While it is true that the return on 52 week T-Bills rose into double digits for three of the 45 years, the interest rate during most of the years was unusually low. Had the pool of Social Security funds been invested in high-quality corporate bonds, the interest rate would have been about 2% higher than the return on T-Bills. Bonds provide a fixed rate of return and should not be confused with the two-way stock market.

Taking into account the variation in life expectancy among males and females, white, African Americans, and other races, in 1995 the average 65 year old person was expected to live about 15 years. Assuming the investment pool operates in a manner similar to an insurance annuity, John Doe's equity of $170,169 would provide him an income of around $1440 for the remainder of his life. This payout is based on the pool earning 6% on funds set aside for distribution to retirees. This payout schedule is shown in Table D.3. If the funds had been invested in high grade corporate bonds, the monthly benefits would have been significantly more.

For comparative purposes, John Doe's history of earnings was furnished to Social Security. Figuring Social Security payments are quite complicated because they adjust annual earnings for inflation and average a number of the annual earnings, etc. However, Social Security generously took the time and effort to determine the amount of Doe's monthly benefits[2]. A comment aside, the author wishes to point out he found the Social Security personnel very courteous and cooperative. Their calculations showed John Doe would receive benefits of about $1078 per month.

Clearly Social Security is no bargain for John Doe, or anyone having a similar work history. The benefits fall about 30% or more short of what an enrollee might receive if allowed to contribute to a retirement pool invested in government securities. If you believe a Social Security participant is entitled to earn interest on his or her contributions during the many years prior to receiving benefits, then it's really beside the point as to whether Social Security was or was not intended as a retirement fund. A participant with a work history similar to John Doe's is entitled to all the Social Security benefits he or she gets and then some.

Whether or not the John Doe example represents the millions of participants in the program can only be answered by an extensive Social Security Administration study of the records. However, John

Doe is probably a good example of how many persons view their relationship with Social Security.

In the John Doe example, his 1979 annual earnings fell below the maximum income on which Social Security collects taxes. The point, for someone whose annual earning's always exceeded the maximum taxed by Social Security, the contributions would have been more than Doe's; the total contributions would have amounted to $96,486.22 versus $64,223.94 to be exact. Invested in 52 week T-Bills, these maximum contributions would have grown to $206,253.05 during the 45 year period. This nest egg would provide an income for life of about $1850 per month. At the time this analysis was made, Social Security's payments for an individual reached a maximum of $1199 per month for a retiree at 65. Again, in this instance, Social Security benefits do not equal what an enrollee is entitled to receive.

For retirees at 66, 67, 68, 69 and 70, Social Security benefits maxed out at about $1259, $1322, $1388, $1457 and $1530 respectively; again at the time this analysis was made. But remember, the retiree's life expectancy has shortened. Also recall that an annuity grows the fastest during the last years because the principal is larger. Hence, an annuity would probably be paying at least $1000 more per month than Social Security for a person retiring at age 70.

Because the assets of Social Security are limited, incoming receipts must be quickly used to pay benefits. Hence, assets have little time to earn interest. To maximize earnings the government uses the accounting practice of first-in, last-out. The procedure increases interest earnings when Social Security is temporarily flush with funds, as in the year 2001. However, in the years ahead the number of people qualified to receive Social Security benefits is projected to swell. Further, recipients are expected to live longer. These two factors increase the depletion rate of Social Security funds. Without a temporary increase in assets, the interest earned disappears and taxes must be increased or some other means must be used to increase Social Security receipts. Since initiated, Social Security taxes have been raised 20 times and are up by a factor greater than seven, from 2% to 15.30%. Further, the maximum salary upon which Social

Security taxes are collected is up from $3,000 to $80,400 and has been raised annually since 1971.

References:

1) AARP Focus on Social Security, July-August 1995, "Modern Maturity"

2) March 24, 1995 FAX from Tom Margenao at the Social Security Administration Press Office, Baltimore, MD to R. Earl Hadady

Social Security Tax Schedule

Year	Salary		Employed Person: Taxes on Earnings, Employee & Employer, Each					Employed Person: Total Tax Paid By Employee & Employer			Self-employed Person: Taxes on Earnings				
	Base $s (max. taxable salary) (a)	Annual % Change	OASI % (b)	DI % (c)	OASDI %	HI % (d)	Total %	% (e)	OASDI Change	Maximum $s	OASI % (b)	DI % (c)	OASDI %	HI % (d)	Total %
1937-49	3,000		1.000	0.000	1.000	0.00	1.000	2.00		60.00	0.0000	0.0000	0.000	0.000	0.000
1950	3,000	0.0	1.500	0.000	1.500	0.00	1.500	3.00	1	90.00	0.0000	0.0000	0.000	0.000	0.000
1951-53	3,600	20.0	1.500	0.000	1.500	0.00	1.500	3.00		108.00	1.8750	0.3750	2.250	0.000	2.250
1954	3,600	0.0	2.000	0.000	2.000	0.00	2.000	4.00	2	144.00	2.6250	0.3750	3.000	0.000	3.000
1955-56	4,200	16.7	2.000	0.000	2.000	0.00	2.000	4.00		168.00	2.6250	0.3750	3.000	0.000	3.000
1957-58	4,200	0.0	2.000	0.250	2.250	0.00	2.250	4.50	3	189.00	3.0000	0.3750	3.375	0.000	3.375
1959	4,800	14.3	2.250	0.250	2.500	0.00	2.500	5.00	4	240.00	3.3750	0.3750	3.750	0.000	3.750
1960-61	4,800	0.0	2.750	0.250	3.000	0.00	3.000	6.00	5	288.00	4.1250	0.3750	4.500	0.000	4.500
1962	4,800	0.0	2.875	0.250	3.125	0.00	3.125	6.25	6	300.00	4.3250	0.3750	4.700	0.000	4.700
1963-65	4,800	0.0	3.375	0.250	3.625	0.00	3.625	7.25	7	348.00	5.0250	0.3750	5.400	0.000	5.400
1966	6,600	37.5	3.500	0.350	3.850	0.35	4.200	8.40	8	554.40	5.2750	0.5250	5.800	0.350	6.150
1967	6,600	0.0	3.550	0.350	3.900	0.50	4.400	8.80	9	580.80	5.3750	0.5250	5.900	0.500	6.400
1968	7,800	18.2	3.325	0.475	3.800	0.60	4.400	8.80		686.40	5.0875	0.7125	5.800	0.600	6.400
1969	7,800	0.0	3.725	0.475	4.200	0.60	4.800	9.16	10	714.48	5.5875	0.7125	6.300	0.600	6.900
1970	7,800	0.0	3.650	0.550	4.200	0.60	4.800	9.16		714.48	5.4750	0.8250	6.300	0.600	6.900
1971	7,800	0.0	4.050	0.550	4.600	0.60	5.200	10.40	11	811.20	6.0750	0.8250	6.900	0.600	7.500
1972	9,000	15.4	4.050	0.550	4.600	0.60	5.200	10.40		936.00	6.0750	0.8250	6.900	0.600	7.500

Table D.1a

Social Security Tax Schedule

Year	Salary		Employed Person								Self-employed Person				
			Taxes on Earnings					Total Tax Paid By			Taxes on Earnings				
			Employee & Employer, Each					Employee & Employer							
	Base $s (max. taxable salary)	Annual % Change	OASI %	DI %	OASDI %	HI %	Total %	%	OASDI Change	Maximum $s	OASI %	DI %	OASDI %	HI %	Total %
	(a)		(b)	(c)		(d)		(e)			(b)	(c)		(d)	
1973	10,800	20.0	4.300	0.550	4.850	1.00	5.850	11.70	12	1,263.60	6.2050	0.7950	7.000	1.000	8.000
1974	13,200	22.2	4.375	0.575	4.950	0.90	5.850	11.70	13	1,544.40	6.1850	0.8150	7.000	0.900	7.900
1975	14,100	6.8	4.375	0.575	4.950	0.90	5.850	11.70		1,649.70	6.1850	0.8150	7.000	0.900	7.900
1976	15,300	8.5	4.375	0.575	4.950	0.90	5.850	11.70		1,790.10	6.1850	0.8150	7.000	0.900	7.900
1977	16,500	7.8	4.375	0.575	4.950	0.90	5.850	11.70		1,930.50	6.1850	0.8150	7.000	0.900	7.900
1978	17,700	7.3	4.275	0.775	5.050	1.00	6.050	12.10	14	2,141.70	6.0100	1.0900	7.100	1.000	8.100
1979	22,900	29.4	4.330	0.750	5.080	1.05	6.130	12.26	15	2,807.54	6.0100	1.0400	7.050	1.050	8.100
1980	25,900	13.1	4.520	0.560	5.080	1.05	6.130	12.26		3,175.34	6.2725	0.7775	7.050	1.050	8.100
1981	29,700	14.7	4.700	0.650	5.350	1.30	6.650	13.30	16	3,950.10	7.0250	0.9750	8.000	1.300	9.300
1982	32,400	9.1	4.575	0.825	5.400	1.30	6.700	13.40	17	4,341.60	6.8125	1.2375	8.050	1.300	9.350
1983	35,700	10.2	4.575	0.825	5.400	1.30	6.700	13.40		4,783.80	7.1125	0.9375	8.050	1.300	9.350
1984	37,800	5.9	5.200	0.500	5.700	1.30	7.000	14.00	18	5,292.00	10.4000	1.0000	11.400	2.600	14.000
1985	39,600	4.8	5.200	0.500	5.700	1.35	7.050	14.10		5,583.60	10.4000	1.0000	11.400	2.700	14.100
1986	42,000	6.1	5.200	0.500	5.700	1.45	7.150	14.30		6,006.00	10.4000	1.0000	11.400	2.900	14.300
1987	43,800	4.3	5.200	0.500	5.700	1.45	7.150	14.30		6,263.40	10.4000	1.0000	11.400	2.900	14.300
1988	45,000	2.7	5.530	0.530	6.060	1.45	7.510	15.02	19	6,759.00	11.0600	1.0600	12.120	2.900	15.020
1989	48,000	6.7	5.530	0.530	6.060	1.45	7.510	15.02		7,209.60	11.0600	1.0600	12.120	2.900	15.020

Table D.1b

Social Security Tax Schedule

Year	Salary		Employed Person: Taxes on Earnings, Employee & Employer, Each					Employed Person: Total Tax Paid By Employee & Employer			Self-employed Person: Taxes on Earnings				
	Base $s (max. taxable salary) (a)	Annual % Change	OASI % (b)	DI % (c)	OASDI %	HI % (d)	Total %	% (e)	OASDI Change	Maximum $s	OASI % (b)	DI % (c)	OASDI %	HI % (d)	Total %
1990	51,300	6.9	5.600	0.600	6.200	1.45	7.650	15.30	20	7,848.90	11.2000	1.2000	12.400	2.900	15.300
1991	53,400	4.1	5.600	0.600	6.200	1.45	7.650	15.30		8,170.20	11.2000	1.2000	12.400	2.900	15.300
1992	55,500	3.9	5.600	0.600	6.200	1.45	7.650	15.30		8,491.50	11.2000	1.2000	12.400	2.900	15.300
1993	57,600	3.8	5.600	0.600	6.200	1.45	7.650	15.30		8,812.80	11.2000	1.2000	12.400	2.900	15.300
1994	60,660	5.3	5.260	0.940	6.200	1.45	7.650	15.30		9,280.98	10.5200	1.8800	12.400	2.900	15.300
1995	61,200	0.9	5.200	0.940	6.140	1.45	7.590	15.18		9,290.16	10.5200	1.8800	12.400	2.900	15.300
1996	62,700	2.5	5.200	0.940	6.140	1.45	7.590	15.18		9,517.86	10.5200	1.8800	12.400	2.900	15.300
1997	65,400	4.3	5.350	0.850	6.200	1.45	7.650	15.30		10,006.20	10.7000	1.7000	12.400	2.900	15.300
1998	68,400	4.6	5.350	0.850	6.200	1.45	7.650	15.30		10,465.20	10.7000	1.7000	12.400	2.900	15.300
1999	72,600	6.1	5.350	0.850	6.200	1.45	7.650	15.30		11,107.80	10.7000	1.7000	12.400	2.900	15.300
2000	76,200	5.0	5.300	0.900	6.200	1.45	7.650	15.30		11,658.60	10.6000	1.8000	12.400	2.900	15.300
2001	80,400	5.5	5.300	0.900	6.200	1.45	7.650	15.30		12,301.20	10.6000	1.8000	12.400	2.900	15.300
2002	84,900	5.6	5.300	0.900	6.200	1.45	7.650	15.30		12,989.70	10.6000	1.8000	12.400	2.900	15.300
2003	87,000	2.5	5.300	0.900	6.200	1.45	7.650	15.30		13,311.00	10.6000	1.8000	12.400	2.900	15.300
2004	87,900	1.0	5.300	0.900	6.200	1.45	7.650	15.30		13,448.70	10.6000	1.8000	12.400	2.900	15.300

(a) Taxes are assessed only on earnings up to and including this amount. (b) Old-age Surviors insurance (c) Disability Insurance
(d) Health Insurance - Medicare, Part A (e) With minor exceptions, self-employed pay both the employer's and employee's taxes.

Table D.1c

John Doe's Social Security Payments

Year	Years Worked	Social Security Deductions: Up to salary of	% of Salary: OASI	DI	Total	John Doe's: Age	Salary per: $s/mo.	$s/year	$s Paid into Social Security: by Doe	by Employer	Total	Year End Earnings: 52 Week T-Bills (%)	Cumulative $s
1950	1	3,000	1.500	0.000	1.500	20	500	6,000	45.00	45.00	90.00	3.00	92.70
1951	2	3,600	1.500	0.000	1.500	21	520	6,240	54.00	54.00	108.00	3.00	206.72
1952	3	3,600	1.500	0.000	1.500	22	541	6,490	54.00	54.00	108.00	3.00	324.16
1953	4	3,600	1.500	0.000	1.500	23	562	6,749	54.00	54.00	108.00	3.00	445.13
1954	5	3,600	2.000	0.000	2.000	24	585	7,019	72.00	72.00	144.00	3.00	606.80
1955	6	4,200	2.000	0.000	2.000	25	608	7,300	84.00	84.00	168.00	3.00	798.05
1956	7	4,200	2.000	0.000	2.000	26	633	7,592	84.00	84.00	168.00	3.00	995.03
1957	8	4,200	2.000	0.250	2.250	27	658	7,896	94.50	94.50	189.00	3.00	1,219.55
1958	9	4,200	2.000	0.250	2.250	28	684	8,211	94.50	94.50	189.00	3.00	1,450.80
1959	10	4,800	2.250	0.250	2.500	29	712	8,540	120.00	120.00	240.00	3.00	1,741.53
1960	11	4,800	2.750	0.250	3.000	30	740	8,881	144.00	144.00	288.00	3.10	2,092.44
1961	12	4,800	2.750	0.250	3.000	31	770	9,237	144.00	144.00	288.00	2.80	2,447.10
1962	13	4,800	2.875	0.250	3.125	32	801	9,606	150.00	150.00	300.00	2.89	2,826.49
1963	14	4,800	3.375	0.250	3.625	33	833	9,990	174.00	174.00	348.00	3.11	3,273.21
1964	15	4,800	3.375	0.250	3.625	34	866	10,390	174.00	174.00	348.00	3.70	3,755.20
1965	16	4,800	3.375	0.250	3.625	35	900	10,806	174.00	174.00	348.00	3.89	4,262.81
1966	17	6,600	3.500	0.350	3.850	36	936	11,238	254.10	254.10	508.20	4.78	4,999.07
1967	18	6,600	3.550	0.350	3.900	37	974	11,687	257.40	257.40	514.80	4.16	5,743.24
1968	19	7,800	3.325	0.475	3.800	38	1,013	12,155	296.40	296.40	592.80	5.68	6,695.93
1969	20	7,800	3.725	0.475	4.200	39	1,053	12,641	327.60	327.60	655.20	6.88	7,856.89
1970	21	7,800	3.650	0.550	4.200	40	1,096	13,147	327.60	327.60	655.20	7.07	9,113.89
1971	22	7,800	4.050	0.550	4.600	41	1,139	13,673	358.80	358.80	717.60	5.33	10,355.51
1972	23	9,000	4.050	0.550	4.600	42	1,185	14,220	414.00	414.00	828.00	4.71	11,710.26

Table D.2a

John Doe's Social Security Payments

Year	Years Worked	Social Security Deductions				John Doe's						Year End Earnings	
		Up to	% of Salary				Salary per		$s Paid into Social Security			52 Week	Cumulative
		salary of	OASI	DI	Total	Age	$s/mo.	$s/year	by Doe	by Employer	Total	T-Bills (%)	$s
1973	24	10,800	4.300	0.550	4.850	43	1,232	14,788	523.80	523.80	1,047.60	7.05	13,657.28
1974	25	13,200	4.375	0.575	4.950	44	1,282	15,380	653.40	653.40	1,306.80	8.16	16,185.15
1975	26	14,100	4.375	0.575	4.950	45	1,333	15,995	697.95	697.95	1,395.90	5.86	18,611.30
1976	27	15,300	4.375	0.575	4.950	46	1,386	16,635	757.35	757.35	1,514.70	6.12	21,357.72
1977	28	16,500	4.375	0.575	4.950	47	1,442	17,300	816.75	816.75	1,633.50	5.14	24,172.96
1978	29	17,700	4.275	0.775	5.050	48	1,499	17,992	893.85	893.85	1,787.70	7.53	27,915.50
1979	30	22,900	4.330	0.750	5.080	49	1,559	18,712	950.56	950.56	1,901.13	8.81	32,443.48
1980	31	25,900	4.520	0.560	5.080	50	1,622	19,460	988.59	988.59	1,977.18	7.54	37,015.97
1981	32	29,700	4.700	0.650	5.350	51	1,687	20,239	1,082.78	1,082.78	2,165.55	13.22	44,361.32
1982	33	32,400	4.575	0.825	5.400	52	1,754	21,048	1,136.61	1,136.61	2,273.22	12.57	52,496.50
1983	34	35,700	4.575	0.825	5.400	53	1,824	21,890	1,182.08	1,182.08	2,364.15	8.87	59,726.79
1984	35	37,800	5.200	0.500	5.700	54	1,897	22,766	1,297.66	1,297.66	2,595.31	10.93	69,133.91
1985	36	39,600	5.200	0.500	5.700	55	1,973	23,677	1,349.56	1,349.56	2,699.12	7.27	77,055.30
1986	37	42,000	5.200	0.500	5.700	56	2,052	24,624	1,403.54	1,403.54	2,807.09	6.32	84,909.69
1987	38	43,800	5.200	0.500	5.700	57	2,134	25,609	1,459.69	1,459.69	2,919.37	6.35	93,406.21
1988	39	45,000	5.530	0.530	6.060	58	2,219	26,633	1,613.95	1,613.95	3,227.91	6.99	103,388.84
1989	40	48,000	5.530	0.530	6.060	59	2,308	27,698	1,678.51	1,678.51	3,357.02	7.84	115,114.74
1990	41	51,300	5.260	0.940	6.200	60	2,401	28,806	1,785.98	1,785.98	3,571.96	7.53	127,623.80
1991	42	53,400	5.260	0.940	6.200	61	2,497	29,958	1,857.42	1,857.42	3,714.84	5.96	139,166.42
1992	43	55,500	5.260	0.940	6.200	62	2,596	31,157	1,931.72	1,931.72	3,863.43	3.98	148,722.44
1993	44	57,600	5.260	0.940	6.200	63	2,700	32,403	2,008.98	2,008.98	4,017.97	3.39	157,918.31
1994	45	60,660	5.260	0.940	6.200	64	2,808	33,699	2,089.34	2,089.34	4,178.69	4.98	170,169.43
							Cumulative total 1950 through 1994			→	64,223.94		170,169.43

Table D.2b

Table D.3a

John Doe Benefits
Invested in 52 Week T-Bills

Had John Doe been able to invest in 52 week T-bills for 45 years vis-à-vis Social Security, at age 65 he would have accumulated $170,169.41 for retirement-see Figure D-2. This accumulation would provide Doe annual benefit of $17,289 for 15 years, his average life expectancy at age 65-see the following spread sheet in Figure D-3b.

As of 1996, Doe's average life expectancy was determined from the following data.

From actuary tables, the average life expectancies at 65 were:

White male	13.7 years
Black male	10.7
White female	18.0
Black female	15.5

The population distribution as of 1996 was:

White	80.3%
Black	12.1
Other	07.6

From the above, the approximate average life expectancy for the total population is:

((13.7+18.0)/2)*0.8 + ((10.7+15.5)/2)*0.2 = 15.3 years

Table D.3b

Year End	Assets: Beginning of year $s	Assets: End of year $s	Earnings @ 6% $s	Benefit Payment $s
Start		170,169		17,289
1	170,169	163,090	10,210	17,289
2	163,090	155,587	9,785	17,289
3	155,587	147,633	9,335	17,289
4	147,633	139,202	8,858	17,289
5	139,202	130,265	8,352	17,289
6	130,265	120,792	7,816	17,289
7	120,792	110,750	7,248	17,289
8	110,750	100,106	6,645	17,289
9	100,106	88,824	6,006	17,289
10	88,824	76,864	5,329	17,289
11	76,864	64,187	4,612	17,289
12	64,187	50,749	3,851	17,289
13	50,749	36,505	3,045	17,289
14	36,505	21,406	2,190	17,289
15	21,406	5,402	1,284	17,289
16	5,402	-11,563	324	17,289
17	-11,563	-29,546	-694	17,289

Federal Employees Retirement Trust Liabilities

At any specified time, such as the end of a fiscal year, the Federal Employees Retirement Trust has these two basic liabilities:

1) **Present Benefits Liability** – the obligation to pay benefits to current beneficiaries until they are deceased
2) **Future Benefits Liability** – the obligation to pay benefits to federal employees who have paid into the trust but as yet are not qualified to receive benefits. At the specified time it is the equivalent of what the employees have paid into the trust plus reasonable earnings.

How these two liabilities may be calculated is covered in Appendix B.

As of the close of the specified year, the assets of the Federal Employees Retirement Trust were more than adequate to meet its **Present Benefits Liability**. However, the funds in the trust were short of meeting the **Future Benefits Liability**. The sum of the two liabilities is shown in the following table.

Unfunded Benefit Liabilities
Dollars are in trillions (1,000 billion = 1 trillion)

Fiscal Year	Present	Future	Total
2001	$ -0.755 t	$ 1.724 t	$ 0.969 t

The preceding dollar numbers were developed from spread sheets which are included in this appendix as Tables E.1 and E.2.

Federal Employees Retirement *Present* Benefits Liability as of 2001

Table E.1 which follows reveals there is **no Present Benefits Liability,** that is funds needed to pay beneficiaries until they are deceased. A **surplus** of **$0.755 trillion** exists as of the close of fiscal year 2001

The **Future Benefits Liability** follows this presentation.

The Present Benefits Liability, Table E.1, assumes:

1) Beneficiaries are equally divided between the sexes.
2) Participants retire at age 57.5 (datum from the government's trust management group) and die at 84.1 (from actuary tables)—an average life expectancy at retirement of 26.6 years (319.2 months).
3) The total monthly benefit payments linearly declines and reaches zero after 26.6 years as beneficiaries become deceased.
4) Cost of living (COLAs) and other adjustments, which would have increased the liabilities, have been ignored
5) Assets earn interest at the rate of 0.5% each month; slightly over 6% per year. Earnings are added to assets each month before benefit payments are deducted.
6) In Table E.1, the first entries under <u>Assets</u> and <u>Benefit Payments</u> are from the September 30th, 2001 issue of the *Monthly Treasury Statement of Receipts and Outlays.* Outlays in the *Treasury Statement* may be an average of the preceding 12 month period to obtain a more representative number for <u>Benefit Payments</u>.

Table E.1

Year	**Month**	**Assets***		**Earnings**	**Benefit**
Fiscal		Beginning of Month ($s m) (1)	End of Month ($s m) (2)	At 0.5% per month ($s m) (3)	**Payments** (end of month) ($s m) (4)
2001	12		554,346		4,018
2002	1	554,346	553,112	2,772	4,005
	2	553,112	551,885	2,766	3,993
	3	551,885	550,664	2,759	3,980
	4	550,664	549,450	2,753	3,968
	5	549,450	548,242	2,747	3,955
	6	548,242	547,041	2,741	3,942
	7	547,041	545,846	2,735	3,930
	8	545,846	544,658	2,729	3,917
	9	544,658	543,477	2,723	3,905
	10	543,477	542,302	2,717	3,892
	11	542,302	541,134	2,712	3,880
	12	541,134	539,973	2,706	3,867
2003	13	539,973	538,818	2,700	3,854
	14	538,818	537,670	2,694	3,842
	306	705,154	708,513	3,526	166
	307	708,513	711,902	3,543	154
	308	711,902	715,321	3,560	141
	309	715,321	718,769	3,577	128
	310	718,769	722,247	3,594	116
	311	722,247	725,755	3,611	103
	312	725,755	729,293	3,629	91
2028	313	729,293	732,861	3,646	78
	314	732,861	736,460	3,664	65
	315	736,460	740,090	3,682	53
	316	740,090	743,750	3,700	40
	317	743,750	747,441	3,719	28
	318	747,441	751,163	3,737	15
	319	751,163	754,916	3,756	3
					-10

Surplus at the end of 319 months

* Calculation basis is: Column (1) + Column (3) - Column (4) = Column (2)

.... Month rows 15 through 305 compressed and hidden

Federal Employees Retirement *Future* Benefits Liability as of 2001

Table E.2 which follows reveals a **Future Benefits Liability** of **$1.724 trillion** as of the close of calendar year 2001—the shortage of funds needed to pay participants who will become beneficiaries in the future.

The Federal Employees Retirement Trust operates on the basis of the government's fiscal year which begins October 1st through September 30th.

For the **Present Benefits Liability** see the preceding pages.

In Table E.2 which follows, calculation of the **Future Benefits Liability** is based on:

1) the procedure and graph in Appendix B.
2) participation in the trust beginning at age 20 and continuing until age 57.5 (the average retirement age was supplied by the U.S. Office of Personnel Management)—a time span of 37 years.
3) data furnished by U.S. Office of Personnel Management.

Table E.2

Year	Receipts		Liability			
			Fractional		Annual	Cumulative
(fiscal)	$s m	Growth	Part	$s m	$s m*	$s m**
1964	2,456					
1965	2,664	8.5%	1/37	72	76	76
1966	2,823	6.0%	2/37	153	162	243
1967	3,094	9.6%	3/37	251	266	523
1968	3,434	11.0%	4/37	371	394	948
1969	3,753	9.3%	5/37	507	538	1,543
1970	4,683	24.8%	6/37	759	805	2,440
▪▪▪▪▪▪						
1986	42,600	6.0%	22/37	25,330	26,850	223,485
1987	43,249	1.5%	23/37	26,885	28,498	265,392
1988	46,284	7.0%	24/37	30,022	31,823	313,139
1989	48,797	5.4%	25/37	32,971	34,949	366,876
1990	52,700	8.0%	26/37	37,032	39,254	428,143
1991	56,800	7.8%	27/37	41,449	43,936	497,767
1992	60,000	5.6%	28/37	45,405	48,130	575,763
1993	62,900	4.8%	29/37	49,300	52,258	662,567
1994	63,800	1.4%	30/37	51,730	54,834	757,155
1995	66,100	3.6%	31/37	55,381	58,704	861,288
1996	67,700	2.4%	32/37	58,551	62,064	975,030
1997	70,400	4.0%	33/37	62,789	66,557	1,100,088
1998	72,800	3.4%	34/37	66,897	70,911	1,237,004
1999	74,500	2.3%	35/37	70,473	74,701	1,385,926
2000	76,000	2.0%	36/37	73,946	78,383	1,547,464
2001	79,146	4.1%	1	79,146	83,895	1,724,207

Future Benefits Liability as of 2001 (→ 1,724,207)

▪▪▪▪▪▪ Year rows 1971 through 1985 compressed and hidden

* Fractional dollars + annual earnings @ 6%

** Prior year + annual earnings @ 6% + current year

Military Retirement Trust Liabilities

At any given time such as the end of a year, the Military Retirement Trust has these two basic liabilities:

1) **Present Benefits Liability** – the obligation to pay benefits to current beneficiaries until they are deceased
2) **Future Benefits Liability** – the obligation to pay benefits to military personnel who as yet are not qualified to receive benefits. At the specified time it is the equivalent of what the government should have paid into the trust on behalf of military personnel plus reasonable earnings.

How these two liabilities may be calculated is covered in Appendix B.

The following is a summary of the two unfunded liabilities as of the close of fiscal year 2001:

Unfunded Benefit Liabilities

Dollars are in trillions (1,000 billion = 1 trillion)

Fiscal Year	Present	Future	Total
2001	$ 0.456 t	$ 0.478 t	$ 0.934 t

The preceding approximate dollar numbers were developed from spread sheets, Table F.1 and Table F.2, which are included in this appendix.

This fund has an unusually long benefit payout period because the average age at which military personnel retire is only 41.3.

Military Retirement *Present* Benefits Liability as of 2001

The spread sheet, Table F.1, which follows reveals a **Present Benefits Liability** of **$0.456 trillion**, as of the close of calendar year 2001—the shortage of funds needed to pay beneficiaries until they are deceased

The **Future Benefits Liability** follows this presentation.

The **Present Benefits Liability,** Table F.1, assumes:

1) Beneficiaries are principally males.
2) Enlisted personnel out number officers by a factor of about 10. Hence, for the purpose of this estimate, officers have been excluded.
3) Participants retire at age 41.3 (datum supplied by the government's trust management group) and die at 78.6 (from actuary tables)—an average life expectancy at retirement of 37.3 years (447.6 months)
4) The total monthly benefit payments linearly declines and reaches zero after 37.3 years as beneficiaries become deceased.
5) Cost of living (COLAs) and other adjustments, which would have increased the liabilities, have been ignored
6) Assets earn interest at the rate of 0.5% each month; slightly over 6% per year. Earnings are added to assets each month before benefit payments are deducted.
7) In Table F.1 the first entries under Assets and Benefit Payments are from the September 30th, 2001 issue of the *Monthly Treasury Statement of Receipts and Outlays.* Outlays may be an average of the preceding 12 month period in the *Treasury Statement* to obtain a more representative number for Benefit Payments.

Table F.1

Year Fiscal	Month	Assets* Beginning of Month $s m (1)	End of Month $s m (2)	Earnings At 0.5% per month $s m (3)	Benefit Payments (end of month) $s m (4)
2001	12		156,978		2,869
2002	1	156,978	154,900	785	2,863
	2	154,900	152,819	775	2,856
	3	152,819	150,733	764	2,850
	4	150,733	148,643	754	2,843
	5	148,643	146,550	743	2,837
	6	146,550	144,452	733	2,831
	7	144,452	142,350	722	2,824
	8	142,350	140,244	712	2,818
	9	140,244	138,134	701	2,811
	10	138,134	136,020	691	2,805
....					
	66	12,711	10,329	64	2,446
	67	10,329	7,941	52	2,440
	68	7,941	5,547	40	2,433
	69	5,547	3,148	28	2,427
	70	3,148	744	16	2,420
	71	744	-1,666 (Assets exhausted)	4	2,414
	72				2,407
2008	73				2,401
....					
	439				55
	440				49
	441				42
	442				36
	443				29
	444				23
2039	445				17
	446				10
	447				4
	448				-3
		Sum of months 71 through 447			455,747
		Present Benefits Liability as of 2001			

* Calculation basis is: Column (1) + Column (3) - Column (4) = Column (2)

.... Month rows 11 - 65 and 74 - 438 are compressed and hidden.

Military Retirement *Future* Benefits Liability as of 2001

Table F.2 which follows reveals a **Future Benefits Liability** of **$0.478 trillion**, as of the close of calendar year 2001—the shortage of funds needed to pay military personnel who will become beneficiaries in the future.

The Military Retirement Trust operates on the basis of the government's fiscal year which begins October 1st through September 30th.

Present Benefits Liability information will be found on the preceding pages of this Appendix.

The following comments apply to the **Future Benefits Liability** spread sheet, Table F.2, on the following page:

1) It is based on the procedure and graph in Appendix B.
2) The **fractional liability** is that portion of the receipts paid into the trust by the government on behalf of military personnel who as of that time were ineligible to receive benefits.
3) It assumes all government obligations prior to 1985 are covered by the payments into the trust after 1985.

Figure F.2

Year	Receipts†		Liability			
			Fractional		Annual	Cumulative
(fiscal)	$s m	Growth	Part	$s m	$s m*	$s m**
1985	0		1/17	0	0	0
1986	26,544		2/17	3,123	3,310	3,310
1987	32,031	20.7%	3/17	5,653	5,992	9,500
1988	33,117	3.4%	4/17	7,792	8,260	18,330
1989	33,994	2.6%	5/17	9,998	10,598	30,028
1990	34,028	0.1%	6/17	12,010	12,730	44,560
1991	35,970	5.7%	7/17	14,811	15,700	62,934
1992	36,500	1.5%	8/17	17,176	18,207	84,917
1993	35,284	-3.3%	9/17	18,680	19,801	109,813
1994	34,860	-1.2%	10/17	20,506	21,736	138,138
1995	34,624	-0.7%	11/17	22,404	23,748	170,174
1996	33,374	-3.6%	12/17	23,558	24,972	205,356
1997	38,173	14.4%	13/17	29,191	30,943	248,620
1998	37,898	-0.7%	14/17	31,210	33,083	296,620
1999	38,227	0.9%	15/17	33,730	35,753	350,170
2000	38,956	1.9%	16/17	36,664	38,864	410,045
2001	40,826	4.8%	1	40,826	43,276	477,923

Future Benefits Liability as of 2001

* Fractional $s liability plus 6% annual interest
** Prior year + 6% annual interest + current year
† Data from Monthly Treasury Statement which differs slightly from the Trust Management Group's data

Railroad Retirement Trust Liabilities

At any specified time, such as the end of a fiscal year, the Railroad Retirement Trust has two basic liabilities:

1) **Present Benefits Liability** – the obligation to pay benefits to current beneficiaries until they are deceased
2) **Future Benefits Liability** – the obligation to pay benefits to railroad workers who as yet are ineligible to receive benefits due to age, etc.—at the specified time it is the equivalent of what railroad workers have paid into the trust plus reasonable earnings.

How these two liabilities may be calculated is covered in Appendix B.

The following is a summary of the two unfunded liabilities as of the close of fiscal year 2001:

Unfunded Benefit Liabilities
Dollars are in trillions (1,000 billion = 1 trillion)

Fiscal Year	Present	Future	Total
2001	$ 0.048 t	$ 0.406 t	$0.454 t

The preceding approximate dollar numbers were developed from spread sheets, Tables G.1 and G2, included in this appendix.

Railroad Retirement *Present* Benefits Liability as of 2001

The spread sheet, Table G.1, which follows reveals a **Present Benefits Liability** of **$48.150 billion**, as of the close of fiscal year 2001—the shortage of funds needed to pay beneficiaries until they are deceased

The **Future Benefits Liability** follows this presentation.

The **Present Benefits Liability** spread sheet, Table G.1, assumes:

1) Beneficiaries are principally males.
2) The typical railroad worker retires at the age of 62.9 (datum supplied by the Railroad Retirement Board) and dies at age 81.2 (from actuary tables)—an average life expectancy at retirement of 18.3 years (219.6 months)
3) The total monthly benefit payments linearly declines and reaches zero after 18.3 years as beneficiaries become deceased.
4) Cost of living (COLAs) and other adjustments, which would have increased the liabilities, have been ignored
5) Assets earn interest at the rate of 0.5% each month; slightly over 6% per year. Earnings are added to assets each month before benefit payments are deducted.
6) In the spread sheet, Table G.2, the first entries under Assets and Benefit Payments are from the September 30th, 2001 issue of the *Monthly Treasury Statement of Receipts and Outlays.* Outlays in the *Treasury Statement* may be an average of the preceding 12 month period to obtain a more representative number for Benefit Payments.

Figure G.1

Year Fiscal	Month	Assets* Beginning of Month ($s m) (1)	End of Month ($s m) (2)	Earnings At 0.5% per month ($s m) (3)	Benefit Payments (end of month) ($s m) (4)
2001	12		26,865		714
2002	1	26,865	26,289	134	711
	2	26,289	25,713	131	707
	3	25,713	25,137	129	704
	4	25,137	24,562	126	701
	5	24,562	23,987	123	698
	6	23,987	23,412	120	694
	7	23,412	22,838	117	691
	8	22,838	22,264	114	688
	9	22,264	21,691	111	685
	10	21,691	21,118	108	681
	11	21,118	20,545	106	678
	42	3,555	2,995	18	577
	43	2,995	2,436	15	574
	44	2,436	1,877	12	571
	45	1,877	1,319	9	568
	46	1,319	761	7	564
	47	761	204	4	561
	48	204	-353	1	558
2006	49	-353	-909	-2	555
	212				25
	213				21
	214				18
	215				15
	216				12
2020	217				8
	218				5
	219				2
	220				-1
	221				-5

Assets exhauted

Sum of months 48 through 219 → 48,150

Present Benefits Liability as of 2001

* Calculation basis is: Column (1) + Column (3) - Column (4) = Column (2)

•••••••• Month rows 12-41 and 50-211 are compressed and hidden

Railroad Retirement *Future* Benefits Liability as of 2001

The spread sheet, Table G.2, which follows reveals a **Future Benefits Liability** of **$0.406 trillion**, as of the close of calendar year 2001—the shortage of funds needed to pay participants who will become beneficiaries in the future.

The Railroad Retirement Trust operates on the basis of the government's fiscal year which begins October 1st through September 30th.

Present Benefits Liability information will be found on preceding pages.

The following comments apply to the **Future Benefits Liability** spread sheet, Figure G.2, on the following page:

1) It is based on the procedure and graph in Appendix B.
2) It assumes participants begin paying into the trust at age 20 and continue until age 62.9 (the average retirement age was supplied by the Railroad Retirement Board)—a time span of 42.9 years.
3) The **fractional liability** is that portion of the receipts paid into the trust by participants who as of that time were ineligible to receive benefits—see the procedure and graph in Appendix B.

Figure G.2

Year	**Receipts**		**Liability**			
			Fractional		Annual	Cumulative
(fiscal)	$s m	Growth	Part	$s m	$s m*	$s m**
1958	924					
1959	990	7.1%	1/43	23	24	24
1960	1,035	4.6%	2/43	48	51	77
1961	1,023	-1.2%	3/43	71	76	157
1962	1,047	2.3%	4/43	97	103	270
1963	1,115	6.6%	5/43	130	137	423
1964	1,170	4.9%	6/43	163	173	622
1965	1,259	7.6%	7/43	205	217	876
1966	1,324	5.2%	8/43	246	261	1,190
1967	1,522	15.0%	9/43	319	338	1,599
1968	1,462	-4.0%	10/43	340	360	2,056
1969	1,654	13.1%	11/43	423	449	2,628
1970	1,869	13.0%	12/43	522	553	3,338
1971	1,943	3.9%	13/43	587	623	4,161
1972	2,114	8.8%	14/43	688	730	5,141
1973	2,318	9.6%	15/43	809	857	6,306
1974	2,711	16.9%	16/43	1,009	1,069	7,753
1990	11,105	1.9%	32/43	8,264	8,760	121,123
1991	11,830	6.5%	33/43	9,079	9,624	138,014
1992	12,313	4.1%	34/43	9,736	10,320	156,615
1993	11,796	-4.2%	35/43	9,602	10,178	176,190
1994	12,335	4.6%	36/43	10,327	10,947	197,708
1995	13,125	6.4%	37/43	11,293	11,971	221,541
1996	12,423	-5.3%	38/43	10,978	11,637	246,470
1997	13,053	5.1%	39/43	11,839	12,549	273,807
1998	13,979	7.1%	40/43	13,004	13,784	304,020
1999	12,430	-11.1%	41/43	11,852	12,563	334,824
2000	13,511	8.7%	42/43	13,197	13,989	368,902
2001	13,745	1.7%	1	13,745	14,570	405,606

Future Benefits Liability as of 2001 →

* Fractional $s liability plus 6% annual interest
** Prior year + 6% annual interest + current year
† Data from Monthly Treasury Statement which differs slightly from the Railroad Retirement Board's data.

OVERVIEW OF PRESIDENT REAGAN'S GRACE COMMISSION

On February 18, 1982, President Ronald Reagan announced the formation of the President's Private Sector Survey on Cost Control. To head up the survey, the President named J. Peter Grace, chairman and chief executive officer of W. R. Grace and Company, a multi-billion-dollar corporation with worldwide, diversified interests in chemicals, natural resources, consumer services, and other areas.

The survey, which became known as the Grace Commission, was formally launched on June 30, 1982, when the President issued Executive Order 12369 to establish an Executive Committee and specify its mandate. Members of the Executive Committee served without pay and the project was funded, staffed and equipped, to the extent practicable and permitted by law, by the private sector without cost to the federal government. The time, money, and materials required to conduct the survey was estimated at more than $75 million.

The Grace Commission provided a rigorous evaluation of government operations by establishing 36 separate task forces, chaired by 131 highly respected individuals from corporate, academic, and labor positions throughout the nation. The entire staff consisted of over 2,000 volunteers. The task forces were titled:

1) Department of Agriculture
2) Department of the Air Force
3) Department of the Army
4) Automated Data Processing/Office Automation
5) Boards/Commissions - Banking
6) Boards/Commissions - Business
7) Department of Commerce

8) Office of the Secretary of Defense
9) Department of Education
10) Department of Energy
11) Environmental Protection Agency, Small Business Administration, and Federal Emergency Management Agency
12) Federal Construction Management
13) Federal Feeding
14) Federal Hospital Management
15) Federal Management Systems
16) Financial Asset Management
17) Department of Health and Human Services: Department Management, Office of Human Development Services, Action —
18) Department of Health and Human Services: The Public Health Service and the Health Care Financing Administration —
19) Department of Health and Human Services: Social Security Administration —
20) Housing and Urban Development
21) Department of the Interior
22) Department of Justice
23) Department of Labor
24) Land/Facilities/Personal Property
25) Low Income Standards and Benefits
26) Department of the Navy
27) Personnel Management
28) Privatization
29) Procurement/Contracts/Inventory
30) Real Property Management
31) Research and Development
32) Department of State/Agency for International Develop ment/United States Information Agency
33) Department of Transportation
34) Department of the Treasury
35) User Charges
36) Veterans Administration

The report complied by the commission provided 2,478 separate and carefully-projected recommendations. If fully implemented, the cost-cutting and revenue-enhancing proposals described in 36 task-force reports and 11 special subject studies, was projected to slash the federal deficit by $424 billion over a three-year period—some $141 billion on an annual basis. The Commission's final report, marking the completion of its assignment, was submitted to the President on January 16, 1984. Upon review of the report, President Reagan promised the Commission "not just talk, but aggressive action on your recommendations."

A more comprehensive report of the Grace Commission study will be found in *A Taxpayer Survey of The Grace Commission Report*[1].

Now almost two decades later, these conclusions can be drawn about the Grace Commission and its work:

1. The recommendations provided were excellent. There never was a serious rebuttal to any of its findings or suggestions.

2. Members of the Grace Commission, and in particular the now deceased J. Peter Grace, deserved high praise for showing quantitatively in hard numbers that the U.S. public debt problem was solvable at that time.

3. The Commission's report and recommendations for all practical purposes was ignored by Congress—they have been known to take a dim view of anything originated outside their provenance. Consequently, the effects of the report were zilch. The deficits continued to grow unabated.

4. The commission put in quantitative terms what was already known qualitatively, that is, a general and significant reduction in government spending could

be achieved which would solve the public debt problem.

5. No one found an answer to the crux of the problem—an intractable Congress.

References:

1) William R. Kennedy, Jr. and Robert W. Lee, *A Taxpayer Survey of The Grace Commission Report* (Ottawa, IL, Green Hill Publishers, 1984)

References

Author's website: **earl-hadady.com**

U. S. Treasury Department
Library: Judy Lim-Sharp
Phone: 202 622-0990 Fax: 202 622-0018
Trust funds: Ruth Williams Phone: 202 874-9880
Publication: "Monthly Treasury Statement of Receipts and Outlays of the United States Government"
Web Site: www.fms.treas.gov
www.publicdebt.treas.gov
Washington, DC

Social Security
Press Office
Mark Hinkle, Public Affairs Specialist
Phone: 410 965-8904
Fax: 410 966-9973
Web Site: www.ssa.gov
Baltimore, MD

Federal Employees Retirement Fund
Management Information Branch
Retirement and Insurance Service
Chris Brown
Email: CGBrown@opm.gov
Washington, DC

Military Retirement Fund

Department of Defense
Chris Doyle, Chief Actuary
Phone: 703 696-7407
Fax 703 696-4110
Washington, DC

Railroad Retirement Fund

U.S. Railroad Retirement Board
Jim Metlicka, Office of Public Affairs
Phone: 312 751-4766
Fax: 312 751-7154
844 North Rush Street
Chicago, IL 60611-2092

Index

a

b

c

d

e

f

g

h

i

k

l

m

n

o

p

r

S

About the Author

R. Earl Hadady—former publisher and editor of the *Bullish Consensus*, a futures advisory service continuously published since 1964—a service based on Contrary Opinion theory of which Hadady is considered a national authority. He is the author of five books on the futures market, one of which has been translated into Chinese. Articles of his have appeared in Barron's and other major business publications. Hadady also appeared weekly on the Los Angeles TV business station before it was moved to NY.

Hadady also authored, *How Sick is Uncle Sam?,* a book dealing with current and future problems facing the U.S. The book was widely acclaimed by notables including Lee Iacocca, J. Peter Grace, the Governor of Colorado and others.

Treason in High Places was his first suspense novel.

For more bio info see website **earl-hadady.com**

Printed in the United States
27640LVS00002B/40

9 781418 449216